The Inspiration Machine

The Inspiration Machine

Computational Creativity in Poetry and Jazz

EITAN Y. WILF

The University of Chicago Press
Chicago and London

The University of Chicago Press, Chicago 60637
The University of Chicago Press, Ltd., London
© 2023 by The University of Chicago
All rights reserved. No part of this book may be used or reproduced in any manner whatsoever without written permission, except in the case of brief quotations in critical articles and reviews. For more information, contact the University of Chicago Press, 1427 E. 60th St., Chicago, IL 60637.
Published 2023
Printed and bound by CPI Group (UK) Ltd, Croydon, CR0 4YY

32 31 30 29 28 27 26 25 24 23   1 2 3 4 5

ISBN-13: 978-0-226-82831-2 (cloth)
ISBN-13: 978-0-226-82833-6 (paper)
ISBN-13: 978-0-226-82832-9 (e-book)
DOI: https://doi.org/10.7208/chicago/9780226828329.001.0001

Library of Congress Cataloging-in-Publication Data

Names: Wilf, Eitan Y., author.
Title: The inspiration machine : computational creativity in poetry and jazz / Eitan Y. Wilf.
Other titles: Computational creativity in poetry and jazz
Description: Chicago ; London : The University of Chicago Press, 2023. | Includes bibliographical references and index.
Identifiers: LCCN 2023001725 | ISBN 9780226828312 (cloth) | ISBN 9780226828336 (paperback) | ISBN 9780226828329 (ebook)
Subjects: LCSH: Creation (Literary, artistic, etc.) | Creative ability. | Human-computer interaction. | Computer composition (Music) | Computer poetry.
Classification: LCC BF411 .W55 2023 | DDC 153.3/5—dc23/eng/20230302
LC record available at https://lccn.loc.gov/2023001725

♾ This paper meets the requirements of ANSI/NISO Z39.48-1992 (Permanence of Paper).

To Henia and Giliad

Contents

Acknowledgments   ix

Introduction: Toward an Anthropology of Computational Creativity   1

PART I Jazz: Mimicry, Originality, Sociality

1. "I Prefer Playing with It to Playing with Most People":
The Computer as a Musical Conversation Partner   35

2. An Island of Interactivity in an Ocean of Nonreactivity:
The Trade-Offs of a Made-to-Order Artificial Musical World   59

3. "A Device That Would Generate New Musical Ideas":
The Computer as a Source of Musical Inspiration   84

4. Separating Noise from Signal: The Ethnomethodological Uncanny
as Aesthetic Pleasure in Human-Machine Interaction   112

PART II Poetry: Indeterminacy, Potentiality, Intentionality

5. Computer-Generated Poetry and Some of Its Aesthetic
and Technical Dimensions   139

6. "I Randomize, Therefore I Think": Computational Indeterminacy and
the Tensions of American Liberal Subjectivity   159

7. Analog Precursors and Their Digital Logical End: The Oulipo   181

8. Crosscurrents and Opposing Perspectives   196

Conclusion: Neither Our Doom nor Our Salvation: Open-Ended
Digital Systems and Cultural Critique   219

Notes   227
References   241
Index   255

Acknowledgments

My deepest gratitude goes first and foremost to my interlocutors in the different ethnographic contexts I explore in this book, who generously shared with me their practices and thoughts and who patiently endured my questions and presence. I have tried to do justice to their respective cultural orders by showing these orders in all of their complexity. In the context of the anthropologist's own cultural order or interpretative framework, it is this complexity that makes cultural orders not only interesting, in the abstract, theoretical sense, but also beautiful, in the most concrete and humane sense.

I am grateful to the people who engaged with some of the arguments I develop in this book when I presented them in different venues, which include the annual meetings of the American Anthropological Association and of the Society for Linguistic Anthropology, the "Ideologies of Design" workshop in the department of anthropology at the University of Chicago, and the Humanities Center at the University of Pittsburgh. Of the many people who provided me with helpful feedback in these venues and in conversations that took place elsewhere, I would like to mention especially Don Brenneis, E. Summerson Carr, Alessandro Duranti, Steven Feld, Susan Gal, Ilana Gershon, Miyako Inoue, Paul Kockelman, Michael Lempert, Paul Manning, David Marshall, Keith Murphy, Constantine Nakassis, Teri Silvio, Michael Silverstein, Jürgen Spitzmüller, and Lucy Suchman.

Mary Al-Sayed, my editor at the University of Chicago Press, should be thanked for taking head-on the unenviable task of enlisting reviewers for this book during the COVID-19 pandemic, when reviewing book manuscripts was the last thing on people's minds. The anonymous reviewers who were eventually enlisted greatly helped make this book's arguments stronger.

Related versions of certain portions of this book have appeared elsewhere as follows: portions of the introduction and of chapters 3 and 4 appeared in "Separating Noise from Signal: The Ethnomethodological Uncanny as Aesthetic Pleasure in Human-Machine Interaction in the United States," *American Ethnologist* 46(2):202–213; "Toward an Anthropology of Computer-Mediated, Algorithmic Forms of Sociality," *Current Anthropology* 54(6):716–739; "From Media Technologies that Reproduce Seconds to Media Technologies that Reproduce Thirds: A Peircean Perspective on Stylistic Fidelity and Style-Reproducing Computerized Algorithms," *Signs and Society* 2(1):185–211; and "Sociable Robots, Jazz Music, and Divination: Contingency as a Cultural Resource for Negotiating Problems of Intentionality," *American Ethnologist* 40(4):605–618. Portions of chapter 6 have appeared in "'I randomize, therefore I think': Computational Indeterminacy and the Tensions of Liberal Subjectivity Among Writers of Computer-generated Poetry in the United States," *American Anthropologist* 50(1):90–102.

I wrote the bulk of this book during the COVID-19 pandemic, a period that produced its own potentialities and limitations. At the same time that it seemed to provide me with much-needed time for reflection and writing by putting everyday life on hold, it created new emotional and practical challenges that threatened to undermine the conditions of possibility and the very justification for pursuing theoretical writing to begin with. I would not have been able to start, continue, and complete the writing of this book during that period had it not been for the sustained support of Anneke Beerkens, my longtime and very organic source of inspiration.

INTRODUCTION

Toward an Anthropology of Computational Creativity

"How Much 'Miles' Will You Have in Your Cocktail, Sir?"

It was a warm day in late August, but inside the lab in an institute of technology in the United States, powerful air conditioners maintained a cool atmosphere. I was still fiddling with the video camera when James, one of the lab directors, entered the room.[1] He nodded to me quickly and then sat down in front of the electric keyboard. Syrus, a robot, was already situated behind the marimba, ready to play, its four arms—each equipped with two mallets—placed in different positions along the marimba. Matt, a member of the research team, sat behind two computer monitors and waited for James's instructions. James turned to Matt and said: "OK, let's do *Yardbird Suite*. Syrus is going to play the head, right?"[2] Matt looked at James from behind the monitors and said, "Yes. And at the end of the head you want Syrus to trade fours with you?"[3] James, playing some quick phrases on the keyboard, replied, "Yes, then trade fours. Does Syrus have a certain amount of Miles, Coltrane, and—I think Syrus has Miles, Coltrane, and You as third, third, third, right?"[4] "Yes," Matt answered, looking at one of the monitors, "but this looks like—because for this project it has a 'You slider,' a 'Charlie Parker slider,' and a—" "Parker, not Coltrane?" James interrupted him with surprise. He looked at Syrus for a few seconds and then said with a smile, "OK. Let it be a third Parker—there can be nothing wrong with having Parker in our mix, right?" Observing this conversation, I agreed wholeheartedly, as did the three students who sat next to me and who, like me, seemed to be curious to know what a robot improvising in a statistical "mix" of the styles of such legendary jazz players as Charlie Parker, John Coltrane, Miles Davis, and the player who happens to play with Syrus on the electric keyboard would sound like. Just before Syrus and James began playing, the student sitting next to me turned to his two friends and asked, laughing and putting on a British accent: "How much 'Miles' will you have in your cocktail, sir?"

*

On the first day of a weeklong workshop on computer-generated poetry, which took place in New York City, Jason, one of the workshop facilitators, had just finished reading an "output text" of a poetry generator that he had created. The code he wrote instructs the computer to randomly slot a preset word list into a number of syntactic templates and thus create brief impressions of narrational meaning. Jason then showed the code to the workshop participants on a large television screen and said:

> As you can see, it's very straightforward. You have a beginning—there are seven choices [of nouns] for the beginning . . . and then there's a verb, and then there are two pronouns, which recur later on. And at the end there are just a few choices. So this is very simple. . . . It's very straightforward. There is very little in the way of generation. The way I think about it—if we were talking about it in the framework of something like artificial intelligence—Herb Simon has this idea of the ant and the beach.[5] You see this ant that's crawling along this very convoluted path, this elaborate path on the beach, and if you know a little bit about the neurophysiology of ants, this path can't be the result of the ant's brain. The ant doesn't have the neurons for it to be doing something this complex. What's the answer to this riddle? Well, the complexity is not in the ant. The ant is following a pheromone trail that's on the beach, so it's the beach and not the ant that has the complexity. This here [the program] is the ant. It's a very simple program. Is there a beach? Is there something that is interesting?

Chris, another workshop facilitator, said, "The fact that such a simple compilation produces all these relations [between the different characters] is interesting." Gregory, a participant, commented, "You [Jason] chose all these words that imply a power structure [between the characters]." Jennifer, another participant, immediately asked, "Are you selecting the words?" Jason said, "I'm not doing anything. My computer is doing everything by itself." Jennifer clarified: "In the Python [code], I mean." "Yeah," said Jason, "I just showed you the program. . . . I wrote the text because I wrote the program, including all the words [that the program uses]." Grant, a participant, said, "I think the beach is the reader. The computer only needs to do the functional part of the language [i.e., slotting the words] more correctly. But the interpretation is outside the computer." Jason then said, "I think of myself as just the ant maker here. I think the beach is a much more complex terrain that includes the reader and the reader's individuality, context, culture, and associations." Shiv, a participant, interjected, "Still, there's something really strange when you know that the construction is made by a computer, because a computer does not have an intention, a precise intention, and this looks quite intentional."

*

This book is an ethnographic study of the ways in which digital computation is impacting notions, practices, and normative ideals of creativity. In different forms of art—from poetry and literature to music and the visual arts—a growing number of artists are taking advantage of digital technologies to transform their creative practices. They do not use these technologies merely to execute their preconceived ideas. Rather, they delegate different degrees of agentive control and artistic decision-making to those technologies to achieve a wide range of goals. Some artists argue that the productive capacities of digital technologies allow them to expand and deepen their own existing creative capacities and even to explore new creative worlds that can challenge their habitual modes of creativity. Other artists point to the democratizing potential of making art-producing digital technologies widely available to the public, as well as to the capacity of such technologies to demystify long-held ideologies about creative genius as the property of unique individuals who work in isolation. Across these different strands, digital technologies are approached and used as active participants in the creative act.

For example, Syrus, the marimba-playing robot, was designed to "imitate" and then "mix" the improvisation styles of different well-known jazz masters and thus play in hitherto new and unheard-of styles.[6] Its computerized algorithms perform a statistical analysis of databases that consist of files of different masters' recorded improvisations. In actual playing sessions, these algorithms instruct Syrus what to play based on this analysis. During the specific session in the lab with whose description I opened this book, Syrus was configured to improvise in a style that is "a mix" of 33.3 percent the style of Miles Davis, 33.3 percent the style of Charlie Parker, and 33.3 percent the style of the player improvising with Syrus on the electric keyboard—in this case, James—whose style Syrus is designed to learn in real time because the electric keyboard is connected via a digital interface to the computer that controls Syrus. It is possible to change the proportions of these different styles in Syrus's playing via sliders on the software interface ("a You slider, a Charlie Parker slider"). Thus, if one wants "more Miles Davis" in Syrus's improvisation, one can manipulate the slider and achieve, for example, a "mix" of 70 percent the style of Miles Davis, 20 percent the style of Charlie Parker, and 10 percent the style of the keyboard player improvising with Syrus.

Similarly, in writing computer-generated poetry, a poet writes a computer program whose execution by the computer results in output texts. Such programs, also known as poetry generators or "works of text generation," typically "transform or reorder one set of base texts of language (word lists, syllables, or preexisting texts) into another form" (Funkhouser 2007:36). Many poetry generators incorporate randomness into their architecture by means of the programming language's random function or by virtue of the reader's input via an interface, while other poetry generators are designed to produce deterministic yet seemingly random sequences of words. As a result, in such cases the poet cannot anticipate in advance the exact nature of his or her poetry generator's potential output texts with complete accuracy. The space occupied by the computer between the poet's intention and the poetic outcome, which raises questions about the creative agency and intentionality that are responsible for this outcome, is a feature rather than a bug—one that is likely to become bigger as artists turn to more complex computational frameworks, such as machine learning.

The active contribution of digital computation to artists' creative practices raises a plethora of questions. How does the use of digital computation as a nonhuman agent (Latour 2005) impact and change the nature of art? What kind of new imaginative spaces does it open up, and what kind of spaces does it make unthinkable? Does the use of digital technologies affect the capacity of art to function as a site of critical thinking in which taken-for-granted assumptions can be challenged? Indeed, given that computerized algorithms often embody cultural prejudices and yet are perceived to be impartial (Noble 2018), does their use in art make the latter a site that reproduces rather than challenges different forms of social injustice and inequality? When artists delegate creative decision-making to technologies whose precise logic eludes them because it is black-boxed, how does their sense of creative intentionality, responsibility, and accountability change as a result (Hill and Irvine 1993)? For example, does their sense of their own creative agency and responsibility shift into that of merely writing code rather than producing the art that results from code's execution? Finally, does digital computation extend the reach of the production and consumption of art, or does it actually restrict it, owing to the fact that digital computation entails new rarefied forms of literacy, skills, evaluation, and, indeed, art, and also necessitates the possession of expensive hardware and software?

On the surface, art seems to be a less consequential domain in which to explore the societal implications of the design and use of digital systems to which different forms of decision-making are delegated than some of the other domains that are now being transformed by these systems, such as criminal justice (Terrio 2019), banking (Stout 2019), health care (Simonite 2019), education (Lutz Fernandez and Lutz 2019), agriculture (Blanchette 2019), communication (Thurman, Lewis, and Kunert 2019), and defense (Masco 2019). However, the design and use of such systems in art might in fact be as consequential as their design and use in these other domains, for a number of reasons.

First, a large share of our fascination with and dread of such digital systems or "roboprocesses" (Gusterson 2019) appears to stem from the fact that they are rapidly becoming sophisticated enough to perform many tasks that we had heretofore thought only humans would be able to perform. In the modern West, creative practice has long been considered to be the epitome of a human skill that defies automation, codification, and rationalization. At least since the European Romantic movement of the mid-eighteenth century, the making of art has been celebrated as humanity's distinguishing feature, the one that sets it apart from other life forms and from machines (Taylor 1989:376). Indeed, prior to that, the capacity for making art was viewed as a divine gift (Abrams 1971:189–190; Murray 1989). Against this ideological backdrop, the specter and appearance of automation in the field of art can legitimize, help spread, and provide the basis for the use of similar digital systems in other domains because it can function as the ultimate proof of concept. Its study is thus a pressing concern.

Second, the cultural contradictions and ethical conundrums that accompany the delegation of human decision-making to digital systems in general, regardless of the specific domain in which this delegation takes place, are likely to be more salient in the field of art both for practitioners and for outsiders because of the unique ideological status of creativity as an irreducibly human faculty, at least in the modern West. The analysis of these contradictions and conundrums in the field of art is thus likely to provide a privileged vantage point from which to study, for example, how the creative "autonomy" that is attributed to these systems is in fact the product of significant human labor and is thus socially constructed and, at the same time, how this social construction nevertheless leads to changes in, and the reevaluation of, the notions of creative agency and practice that informed this construction.

Last, the design and use of roboprocesses in criminal justice, banking, health care, education, agriculture, communication, and defense can be traced to "the emergence and maturation of bureaucratic forms of administration in the nineteenth and twentieth centuries" (Gusterson 2019:4). Such digital systems are thus informed by the impersonal, formalized, standardized, and rational normative dimension of Western modernity's institutions and ideals, in line with Enlightenment-based values. In contrast, the design and use of the digital systems that I explore in this book are informed by an additional key dimension of Western modernity, namely modernity's normative ideals of creative agency, individual expressivity, and uniqueness and originality, which crystallized with the rise of Romanticism. Although these two currents have been equally important in constituting the contemporary modern moment in the West, for most of the last two centuries they have been constructed and understood in opposition to each other (Wilf 2014a:8–13). The design and use of art-producing computerized systems combine these two currents in intimate ways and thus provide an opportunity to study their co-constitution in new empirical settings.[7] The rest of this introduction charts in detail the more specific theoretical and empirical frameworks on which this book builds and to which it seeks to contribute, as well as its methodological foundations.

Mimetic Precursors, Generative Successors

Attempts to automate some dimensions of creative practice predate the rise of digital technologies. For example, eighteenth- and nineteenth-century Europe saw the rising popularity of android automatons (Wise 2007). These were machines, based in watchmaking technology, that were often given the shape of humans and animals playing music, drawing, and writing poetry. One of the most sophisticated of these, known as Maillardet's automaton and made in the form of a child with a pen in its hand, was designed around 1800.[8] After the machine is wound, the "child" can create four highly intricate drawings and write three poems—two in French and one in English. The drawings are beautiful and no doubt surpass what most people would be able to produce if asked to draw a ship, Cupid in a garden, or a Chinese temple—some of the subjects that Maillardet's automaton can draw. Another celebrated example is an automaton designed by David Roentgen to have the form of Marie Antoinette playing intricate pieces on a dulcimer.[9] Roentgen presented it to Louis XVI, Marie Antoinette's husband, in 1784. These and similar mechanical devices and automata that had been built for millennia "provoked questions about ontology, humans and nonhumans, nature and artifice; they challenged the borders separating illusion, reality, and possibility" (Mayor 2018:211; see also Berryman 2007; Landes 2007).

However, as sophisticated as these machines were, they were limited in one profound way: they were not creatively generative. They could only execute a limited number of predetermined scripts—specific poems, specific drawings, specific musical pieces. They did not have the generative capacity to produce new drawings, new paintings, or new musical pieces, or to execute existing pieces that had not been stored in their memory. Although this limitation reflected technological constraints, it was also ideologically motivated. Such machines were designed to entertain their (mostly wealthy white male) human owners by virtue of their mimetic capacities rather than to threaten them with their creative capacities (Taussig 1992:212–235; Wise 2007). In the normative hierarchy of human faculties prevalent in nineteenth-century Europe, imitation ranked lower in comparison with creativity (Wilf 2012). It was associated with women, children, nonwhite humans, and animals. For this reason, most of those automatons were designed to have the appearance of these and similar marginal figures.

One of the poems that Maillardet's automaton can write epitomizes this ideal of limiting the automation of art to the nonthreatening execution of an existing artwork with perfect accuracy and fidelity, guaranteeing that the machine will have no creative autonomy, will only entertain, and will know its place in the normative patriarchal hierarchy:

> Unerring is my hand tho small
> May I not add with truth
> I do my best to please you all
> Encourage then my youth

The poem expresses the automaton's mimetic capacity ("Unerring is my hand"), its desire to follow the mimetic ideal ("May I not add with truth"), its awareness that its sole function is to entertain by imitation rather than to challenge by creative agency ("I do my best to please you all") and, finally, its awareness that, as a child, it is located lower in the social hierarchy and that this marginality is the source of its mimetic capacity ("Unerring is my hand tho small" and "Encourage then my youth").

The digital technologies that I explore in this book are different from such automatons not only by virtue of those technologies' underlying architecture—digital computation rather than watchmaking technology—but also because of the goals that motivate their design and use. First and foremost, rather than mediating and reproducing a limited number of fixed texts in different modalities, these digital technologies are meant to abstract, mediate, or instantiate generative principles that my interlocutors frequently thought of as styles and thus produce new fixed texts in those styles. I deliberately use the term "meant" to make a distinction between the goals, hopes, and aspirations that inform the designers and users of such technologies, and what these technologies accomplish in practice.

As I discuss in chapters 1 and 2, David, a computer scientist who is also a semiprofessional trumpet player, developed a jazz-improvising interactive computerized system whose task is to generate new improvisations in David's own improvisation style. David built this system so he could engage with it in meaningful musical interactions. In designing Syrus, the robot marimba player that I explore in chapters 3 and 4, James, a computer scientist and semi-professional keyboard player, had a different goal in mind: to build an interactive jazz-improvising system that could not only generate new improvisations in existing styles but also "mix" those styles and thus generate new improvisations in as-yet unheard-of styles that might inspire human musicians. Finally, the writers of computer-generated poetry whose work I examine in chapters 5-8 delegate different degrees of agentive control to digital technologies that become a fundamental dimension of their own creative agency and ability to write poetry. Whenever these poets want to write a new computer-generated poem, they write a new code that represents a generative principle whose execution or realization by the computer results in output texts whose exact nature the poets cannot anticipate in advance with complete accuracy.

From Technologies That Mediate Fixed Texts to Technologies That Mediate Styles

Until recently, cultural and linguistic anthropologists have focused primarily on the analysis of media technologies that mediate fixed texts in different modalities such as sound and the visual image. For example, Patrick Eisenlohr has studied devotional practices among Mauritian Muslims who use cassettes, CDs, or MP3 files that mediate emblematic recitations of *na't*, a popular genre of devotional Urdu poetry, by master reciters (Eisenlohr 2010), and I (2012) have studied the pedagogical practices by means of which jazz students in the United States use the same technologies to learn some canonical improvisations of past jazz masters (see also Kunreuther 2010; Spitulnik Vidali 2010; Weidman 2010). In such studies, the analytical focus has been how media ideologies, intertextuality, participation framework, and the politics of mediation shape the ways in which people mobilize technologies that mediate a specific text such as an image of a holy site or a holy figure, or an audiovisual representation of a religious ritual (Meyer 2006).

Two crucial distinguishing features of technologies that are meant to mediate styles separate them from technologies that are meant to mediate fixed texts. Technologies that are meant to mediate styles are designed, first, to synthesize or abstract a style as a generative principle from a corpus of fixed texts, and/or, second, to enact or realize this generative principle by producing new fixed texts in this style. All that a technology that mediates fixed texts can do with respect to style is provide users with the fixed texts from which they, that is, the users, might be able to abstract a style. Here enters the phenomenon of type intertextuality as abstracted by individuals, or an "interpreter's retrospective or recuperative relationship . . . to an internalized notion of a type or genre of discursive event" (Silverstein 2005:9), or "the link" individuals might make not "to isolated utterance, but to generalized or abstracted models of discourse production and reception" (Briggs and Bauman 1992:147). Hence, the claim, made apropos media technologies that mediate fixed texts, that "electronic reproduction of voice facilitates authentic transmission of both the 'correct' poetic text and the appropriate performative style" (Eisenlohr 2009:283) can be held as true as long as it remains clear that the recording itself does not reproduce a performative style or a "typified speech genre" (Eisenlohr 2010:327) or any other genre in whatever modality. Rather, the users of such technologies abstract a style or genre based on the relations they perceive between a number of recordings within the same genre or between the different parts of one recording.

In contrast, a technology that is designed to mediate a style is meant to perform such an abstraction or synthesis from a given corpus of fixed texts and/or to enact or realize such a style by generating new fixed texts in this style. Consequently, its users evaluate it on different terms. This is why the technologies designed to mediate past jazz masters' styles, which I analyze in chapters 3 and 4, are evaluated not in terms of the idea of "sound fidelity" (Sterne 2002), but rather in terms of what I call "stylistic fidelity" (Wilf 2013a)—their ability to abstract a style from a given corpus of existing recorded jazz solos of a past jazz master and to generate an endless series of new solos in the style of that master.

To exemplify and make more concrete the two different kinds of agency represented by these two different kinds of media technologies, I turn to a vignette that I recorded during my previous ethnographic fieldwork in academic jazz programs in the United States (Wilf 2014a)—an ethnographic context that is directly relevant to the two jazz-improvising computerized systems I explore in the first part of this book. The vignette tells the story of a novice player who found himself at a stage of his training in which he could only function as a technology that mediates fixed texts. In classes, clinics, method books, and informal conversations with students, jazz educators stress time and again the importance of transcribing the recorded solos of the past jazz masters and either practicing them in their entirety or dividing them into selected "licks" and phrases and practicing these excerpts in every key. This aspect of jazz training has been widely documented by jazz scholars. Its purpose is to have students incorporate into their playing bodies prototypical features of canonical improvisations.

In an interview I conducted with Henry, a man in his mid-sixties who at the time of my fieldwork was a teacher at Berklee College of Music in Boston, about his own training as an aspiring young student, the potentialities and limitations of this form of training came to the fore. Henry told me the following story:

> [A friend] gave me a Kenny Dorham album to listen to.[10] And one of the things I knew after studying this album was this blues called *Double Clutching*. I learned it note for note with the record. . . . So now I had a five-chorus blues in [the key of] F transcription that I could play. And I knew it inside out, man, and it was swinging. Not many people knew that album, at all, you know. And it was a good mean tempo. So I went to this jam session once, and these guys were good players, better than me, and they didn't know me at all and they said, "So what do you want to do?" So I said, "Eh"—[Henry simulates hesitation] "let's do a blues in F." We played *Billie's Bounce* or some kind of head [melody], you know, and then I played that [Dorham's] solo.[11] Five choruses. Man, when I got done playing that solo these guys were like: "Wow, man, this cat's heavy! Who is this guy?" You know, and I felt good, man. And, eh [Henry starts to laugh], the only thing is, though, that was the only blues solo I could play. And then after I played that solo they said, "Yeah, man, what do you want to do now, man?" And I was like, "Wow, I tell you, man, I think I gotta go home, I'm feeling really shitty." And I pretended that I was coming down with something, you know, like the flu, feeling terrible, and I left. And wow! [Henry says with relief] I got out of there!

The story illustrates the advantages and disadvantages of learning past masters' solos tout court and performing them in their entirety in a live setting. On the one hand, the audience and players took Henry's performance to be a real-time feat of his creativity, because they were not familiar with Dorham's recorded solo and because jazz is an improvisation-based musical genre; it is meant to be a form of composition that occurs on the spur of the moment. The audience and players did not know that Henry's performance was an exact reproduction of one of Kenny Dorham's recorded solos. Henry meticulously learned this solo with the aid of a recording and played it with the live rhythm section with such conviction that everyone thought he was the solo's author rather than its animator (Goffman 1981).[12] The audience and players were impressed because the solo is, indeed, a masterpiece. When recounting another incident in which he did the same thing with one of Freddie Hubbard's solos, Henry told me that "the crowd went crazy."[13]

On the other hand, this practice had severe limitations. Henry's ability to simulate creativity was not generative: it was limited to the scope of those fixed instantiations of past masters' creativity that are the recorded solos of a Kenny Dorham or a Freddie Hubbard. When Henry was asked to display further creativity on other tunes, he had to pretend he was sick and leave the stage. Henry clearly felt his own limitations, which had been the result of practicing the kind of reproduction or mediation that is limited to finite, fixed, and, in this case, existing stretches of musical discourse. Whatever Henry gained by copying the masters' solos in public, he lost when the stretch of copied musical discourse ended and he became fully aware of his own substantial limitations as an improviser. He soon realized that he lacked that principle of generativity that is the essence of style.[14]

In contrast, consider one scene in the movie 'Round Midnight (Eastwood 1986), in which the movie's main character, Dale Bauer, sits down to dinner with a French admirer, Francis, in the latter's apartment. Bauer—played by the jazz legend Dexter Gordon (1923–1990)—is an old, once-famous African American jazz musician who struggles with alcoholism and other demons. As Francis dilutes wine with water to serve to Bauer, the latter says: "All these young kids sound the same, just like they had the same teacher." Francis asks him, "It was you?" Bauer responds, "Yeah, me, and a few others. . . . You know, one night in Brooklyn, this tenor player comes in. And he sits down and he listens [to me playing]. And then he comes up to me and says: 'I play YOU better than YOU'" (Eastwood 1986:00:46:26–00:47:25). What does it mean to play another player better than that player? It means that the intensive study of numerous recordings (as opposed to copying a single recording) of a past master by a student can sometimes result in the student's being able to synthesize and assimilate the master's style of improvisation. This, as opposed to the form of imitation practiced by Henry, entails not the replication of existing improvisational texts of finite duration, but the synthesis and incorporation of a generative principle that allows a student to sound like the master by producing new improvisations in the master's style. The young player who approached Bauer and told him that he played "YOU better than YOU" did not mean that he could play Bauer's earlier recorded solos better than Bauer had played them. Rather, he meant that he had managed to inhabit Bauer's style of improvisation as the generative principle that had been responsible for Bauer's well-known improvisations and, given Bauer's decline, to enact this style better than Bauer when they met in the Brooklyn club.[15]

If Henry exemplifies the media technologies that cultural and linguistic anthropologists have so far tended to investigate, the young player in the movie exemplifies the media technologies that are the focus of this book. Theoretically, the difference between Henry and the young player, and hence between the two kinds of media technologies as their designers and users understand them, can be conceptualized by means of Bourdieu's notion of habitus. Bourdieu argued that the habitus is an individual's propensity, disposition, or generative scheme of action, which stipulates not specific actions in specific situations, but rather the likelihood that certain types of actions will take place in certain types of situations. He rejected objectivist approaches in the social sciences that attempted to explain social behavior as a set of discrete rules of action assigning specific actions to specific situations (Bourdieu 1977:119). Instead, he offered "strategy"— a set of embodied dispositions and generative schemes that are unconsciously acquired in childhood and that account for individuals' ability to improvise in different situations in a consistent way that accords with and reproduces their position in the social structure—as a more accurate explanatory model for practice (Lamaison and Bourdieu 1986).[16] At one point in his *Outline of a Theory of Practice*, Bourdieu defined "the particular stamp marking all the products of the same habitus, whether practices or works," in terms of "style" (Bourdieu 1977:86).[17]

Against this backdrop, Henry felt what it means to lack a specific habitus in a cultural context that requires it, while pretending to have it.[18] Dale Bauer, on the other hand, felt what it means to have your habitus successfully inhabited by another person while you yourself are gradually losing your grip on it. This comparison hints at the potentially radical implications of technologies that are designed to mediate a style, particularly in art. However, these implications are radical only against the backdrop of culturally specific notions of creative agency. Hence, to tease out these implications and thus to understand the desire to design and interact with such technologies, it is necessary to unpack the cultural specificity of the notion of style in modern Western art.

Style as Ideology in Modern Western Art

Modern Western art is informed by normative ideals of creative agency that have a specific intellectual history. These ideals received an emblematic expression in an interview I conducted with Joe, a well-known American jazz saxophonist in his mid-fifties, during a break in a semester-long clinic he gave at Berklee College of Music. Throughout the semester, Joe emphasized time and again the importance of "feeling" with respect to creativity. At one point in the interview, I asked him to articulate in greater detail the relation between creativity and feeling. As he thought over how to respond, Joe lowered his eyes to his saxophone, which was resting in his lap, and said:

> I mean, what is creativity? What is jazz? Jazz is an expression of who you are in your life and how deep you get into music, the elements of music, the theory of music, and how you can express it as a player. You know, what is it to create a solo? I mean, let's go back to the very beginning. The blues and the music— it's all about feeling. Totally. If you copy the way that someone played and play

it, all you do is trying to play what they played. And if they played it with feeling, you can only try to play like their feeling. That's not you. What's your feeling all about? And that will only go so far, too. You can only do that for twenty minutes, maybe two minutes, maybe one minute. Maybe three hours! But a lifetime? Or a complete record? Or a ninety-minute set? Or take after take? That approach is an easy one to teach. It's like if you tried to be a painter and you tried to copy a van Gogh or a Monet, right? What will you end up with? A Monet on a bad day or a van Gogh on a bad day. You know what I'm saying?

The notion of "feeling" that plays such a central role in Joe's commentary is a core dimension of modern normative ideals of creative agency. It has its origins in Sentimentalism, a movement that developed from Protestant Pietism in mid-eighteenth-century Europe (Wilf 2011). Sentimentalism promoted an ideal character type that was based on the subject's susceptibility to his or her tender feelings and to those of other subjects—what one scholar called "the ethic of feeling" (Campbell 1989:138–160)—against the backdrop of and as a reaction to a neoclassicist aesthetics that emphasized the importance of formal, rational, and abstract rules. In this context, "the identification of beauty and the formulation of the good . . . could . . . be ascertained merely by 'trusting to one's feelings'" (151). At the end of the eighteenth century, Sentimentalism evolved into full-blown Romanticism. Drawing on organic metaphors, key Romantic thinkers argued that each individual has his or her own nature or voice or structure of feeling with which he or she must be in touch and to which he or she must remain faithful, and that copying or adhering to external rules or models of feeling amounts to distorting this inner voice: "This notion of an inner voice or impulse, the idea that we find the truth within us, and in particular *in our feelings*—these were the crucial justifying concepts of the Romantic rebellion in its various forms" (Taylor 1989:368–369, emphasis added).[19] Joe argues that each person has his or her unique structure of feeling that is part of his or her identity, and that trying to imitate someone else's structure of feeling can only lead to poor results. This is what he means when he describes copying's results as "a Monet on a bad day or a van Gogh on a bad day." Here "a Monet" and "a van Gogh" designate original styles, where style is a combination of a feeling or quality that is unique to these artists and that is longitudinal, that is, that manifests itself in a consistent manner over time. It is a general principle responsible for the generation of new yet stylistically similar works of art.

Most of the strands of Romantic ideology were premised on the assumption that each individual has a unique and consistent nature or structure of "feeling" as a source of action: "This is the idea which grows in the late eighteenth century that each individual is different and original, and that this originality determines how he or she ought to live" (Taylor 1989:375). This assumption, which is alive and well today, was the basis for the institution of the copyright. Proponents of the copyright argued that the form of an artwork is a reflection of the artist's unique nature and that this form is an inalienable possession that cannot be copied by other people without violating one of the artist's basic rights (Woodmansee 1996:51). What can be passed on to others when they purchase a book, for example, is the physical material of which the book consists and the ideas it contains—but not their specific form, which, because works of art "grow spontaneously from a root, and by implication, unfold their original form from within" (54), is a reflection of the artist's inner nature. The specific form or composition in which these ideas are rendered, then, remains the author's property. Johann Gottlieb Fichte, one of the key architects of these ideas, articulated them in an essay he wrote in 1793:

> Each individual has his own thought process, his own way of forming concepts and connecting them. . . . All that we think we must think according to the analogy of our other habits of thought; and solely through reworking new thoughts after the analogy of our habitual thought processes do we make them our own. . . . Each writer must give his thoughts a certain form, and he can give them no other form than his own because he has no other. But neither can he be willing to hand over this form in making his thoughts public, for no one can appropriate his thoughts without thereby altering their form. This latter thus remains forever his exclusive property. (Quoted in Woodmansee 1996:51–52)

Note that Fichte's argument combines the notion of an original and autochthonous feeling or nature and the notion of the consistency and habituality of this feeling or nature, which one cannot escape. The result is the production of a series of concrete works of art that bear the mark of a single unique style.

Successfully learning and inhabiting someone else's style, then, as the young player did apropos Dale Bauer's style in the movie *'Round Midnight*, is far more subversive, ideologically speaking, than copying his or her specific output, as Henry did apropos Kenny Dorham's solo. Such a feat complicates Joe's argument that the results of copying must be "a Monet on a bad day or a van Gogh on a bad day." It suggests that, in fact, diligent followers can become "a Monet" on a great day or "a van Gogh" on a great day once they decipher and perfect the generative principle or style that is responsible for their oeuvre, as ample cases of successful forgeries demonstrate.[20] If a certain style has its origins in an individual's specific nature or feeling, as Joe argues, and if it is possible for one person to inhabit another person's style, then for all intents and purposes both people come to share and have the same nature or structure of feeling. This possibility generates problems for widespread ideologies that conceptualize creativity in terms of uniqueness, originality, singularity, and inalienability.

Technologies that are designed to mediate styles, especially in the field of modern art, intensely fascinate their developers, users, and observers against this specific ideological backdrop. They reconfigure the participation framework (Goffman 1981) people usually associate with technologies that mediate fixed texts. Whereas a technology that is designed to mediate fixed texts functions mainly as an animator, a technology that is designed to mediate styles functions mainly as an author (and possibly an animator too) that produces novel instantiations of someone else's—that is, the principal's—style. It composes (and sometimes, though not necessarily, executes or animates) the music or poetry or painting on behalf of whoever's style it mediates at a given moment. Hence, when a human musician musically interacts with a system, which learns his or her style in real time and generates music in this style and in response to the musician's playing, the musician and the people who observe him or her interacting with the system often have the impression that they are experiencing and witnessing a kind of real-time proliferation and duplication of various authors who faithfully "speak" or create on behalf of one principal: the human musician (Pachet 2002).

This is not to deny that technologies designed to mediate fixed texts have been an object of fascination, too, as a result of their reconfiguration of the participation framework that people had hitherto been accustomed to. However, the fascination they provoked resulted from their ability to reproduce the animator through the reproduction of the very same material or qualia (such as sounds or colors) of which the original fixed text consisted (Eisenlohr 2010:328–329; Taussig 1992; Weidman 2003; Wilf 2012). This reproduction of qualia has been the focus of efforts to perfect "sound fidelity," "high definition," and similar technical ideals in different modalities (Sterne 2002). In contrast, technologies designed to mediate styles become objects of fascination because they give their users and observers the sensation that they are witnessing the real-time proliferation of authors who faithfully compose on behalf of or in the style of the same person. It is for this reason that such technologies have become the subject of efforts to perfect a new technical ideal: "stylistic fidelity" (Wilf 2013a).

From Mechanical Divination to Art-Producing Computerized Systems

The artists I worked with were motivated to design and use art-producing digital technologies because they believed that these technologies might enrich and transform their own creative agency and intentionality, which faced different kinds of challenges. For example, David built his jazz-improvising system in order to have an improvisational partner, because he felt that the jazz musicians he had been playing with in jam sessions had stopped listening to one another and had instead been satisfied with a solipsistic kind of playing. James, in contrast, designed his system in order to create new improvisation styles in a musical landscape that he felt had been dominated by the same ossified and stale styles for decades. Last, the practitioners of computer-generated poetry I worked with emphasized the opportunity that digital computation gave them to problematize and reconfigure their habitual writing styles.

The very idea of using digital computational technologies to enrich one's creative agency and intentionality, which since the rise of Romanticism have been associated with unpredictability, spontaneity, regeneration, and organic metaphors, might seem to be a contradiction in terms. Indeed, the assumption that creative agency and its products cannot be cultivated in and by highly regimented and formalized contexts and means has been prevalent not only in the popular imagination but in anthropological theory too.[21] In the following chapters I problematize this theoretical tradition by presenting an ethnographic analysis of creative intentionality in the age of digital computation. The design and use of art-producing digital systems conflate creative agency and a regimented and formalized context—that is, digital computation—as never before. Rather than assume that the result would necessarily be a contradiction in terms, I analyze how in different contemporary contexts creative agency and digital computation come to co-constitute each other. I ask what people find to be creative in computation (for example, digital computation's affordance of emergence via randomization, or the execution of nonlinear processes), what they find to be computational in creativity (for example, the fact that many art forms rely on algorithmic processes of composition), how these insights help them to mediate between worlds normally thought to be miles apart from one another, and how their notions of both creative agency and computation are transformed as a result of their practices of mediation.

My interlocutors were especially drawn to digital computation's capacity to generate what I call "contextually-meaningful contingency," which can be used to address different problems of creative agency and intentionality (Wilf 2013b). By "contingency" I mean "uncertainty of occurrence of incidence" or "being open to the play of chance, or of free will" (*Oxford English Dictionary*). Contingency does not necessarily mean complete unpredictability, because it is possible to estimate the chance of occurrence for some contingent events based on various considerations. Thus, the notion of chance, or "the possibility or probability of anything happening: as distinct from a certainty" (ibid.), is also important to the idea of contingency. As will become clear throughout this book, one reason the people I worked with were drawn to digital computation was its capacity to generate output that is contingent—"open to the play of chance, or of free will"—but not completely unpredictable, because it can be constrained by various contextual factors such as the harmonic structure of a jazz tune, the real-time contributions of one's bandmates, or specific poetic dimensions that the poet can define in advance by means of code (see "the dance between accident and design," Hayles 2009:43; see also Taylor 2014:170–171). The fact that the contingent output generated in this way is meaningfully related to such contextual factors is what makes it contextually meaningful.

Contingency has been used as a cultural resource for negotiating problems of agency and intentionality in widely different forms and ethnographic contexts. Consider, for example, the class of practices called mechanical divination, which anthropologists have studied extensively for many decades (Du Bois 1993; Evans-Prichard 1991 [1937]; Moore 1957; Park 1963; Wilce 2001; Zeitlyn 2012). To zoom in even further, consider the divinatory practice known as the poison oracle, as famously analyzed by Evans-Pritchard (1991), in the context of which the client asks a question, a poison is then administered to a fowl, and an answer is divined according to whether the fowl lives or dies. The poison oracle has been discussed as an aleatory mechanism used by the Azande in central Africa to find out who has bewitched them or intends to do so (Du Bois 1993:58). The advantage of this mechanism is that "the meanings arrived at are determined by something other than a volitional, human act" (54). This feature allows the poison oracle to function as a reliable or unbiased "method of revealing what is hidden" (Evans-Pritchard 1991:120), namely, the malicious intentions of one's group members, with which this agonistic cultural context is overflowing. Although Azande individuals do not explicitly treat the poison oracle as a randomizer but rather assign it mystical properties (147–151), the administration of poison to fowls does, in fact, amount to the production of a randomizer. It is for this reason that "as soon as the poison is brought back from its forest home it is tested to discover whether some fowls will live and others die under its influence" (158). A poison that systematically kills all fowls or has no effect on any of them is tantamount to a die that falls on one face time and again: it is not a proper randomizer in that, all things being equal, it is not "governed by or involving equal chances for each of the actual or hypothetical members of a population" (*Oxford English Dictionary*): its outcomes are known in advance. Here, then, as in the contexts I am concerned with in this book, contingency, understood as "uncertainty of occurrence" (ibid.), is mobilized as a cultural resource for negotiating problems of intentionality—in the Azande case, the problem of the agonistic intentionality of one's group members.

However, Evans-Pritchard argued that in consulting the poison oracle, Azande individuals harness contingency as a cultural resource for *reducing* the uncertainty generated by group members' malicious intentions. Similarly, a recent analysis of the use of divination for helping "intentional subjects (the clients) deal with uncertainty" has suggested that even when divination is in fact "redeploying and possibly increasing uncertainty, rather than reducing it," "the uncertainty may be tempered by the possibility of doing something to address the topics raised in the divinatory session" (Zeitlyn 2012:537). The same analytic approach has been prevalent in the long anthropological and folkloristic study of performance, where scholars have argued that individuals might seek to increase or emphasize the existence of contingency but only for the purpose of publicly displaying their competence in taming, reducing, or resolving it in real time (Bauman 1984:129–130; Berliner 1994:374–383; Gioia 1988; Keane 1997; Lord 1960:37–38; Malaby 2002; Rosaldo 1986:134; Schieffelin 1985:721–722).

More generally, anthropologists who have studied the relation between human intentionality and contingency have tended to conceptualize the former as something that is naturally pervaded by the latter. Anthropologists influenced by pragmatism and phenomenology (Csordas 1993; Ingold and Hallam 2007; Jackson 1989; Malaby 2007), for example, have highlighted the open-ended nature of human action in the world and the ways in which our intentions are themselves evolving projects that unfold in time rather than being formulated in advance. They have thus focused on contingency as the very stuff of which intentionality is made. Other anthropologists have highlighted intentionality in relation to the vicissitudes of the external world and of other people's intentions (Becker 1997; Bourdieu 1977; Douglas 1994; Keane 1997; Sahlins 1985). Marshall Sahlins, for example, has emphasized the "double contingency" (Sahlins 1985:145) of agents' subjective interests and the material intractability of the objective world, which accounts for the functional reevaluation of cultural categories whenever they are enacted in practice by intentional agents. Similarly, Webb Keane has pointed to the hazards in which "even the most controlled representational practices" are thoroughly implicated (Keane 1997:xiv), adding that agents' intentions are frequently frustrated or reconfigured because of "the vicissitudes" that result from depending on recognition by others and from the material basis of signs that frequently subverts what actors intend the signs to mean (208).

Similar considerations have led the philosopher Alasdair MacIntyre to suggest general principles that stipulate under what conditions individuals would try to minimize or maximize the contingency of their and other agents' intentions. Among the "general features of social life" that he describes is the following:

> It is necessary, if life is to be meaningful, for us to be able to engage in long-term projects, and this requires predictability; it is necessary, if life is to be meaningful, for us to be in possession of ourselves and not merely to be the creations of other people's projects, intentions and desires, and this requires unpredictability. We are thus involved in a world in which we are simultaneously trying to render the rest of society predictable and ourselves unpredictable, to devise generalizations which will capture the behavior of others and to cast our own behavior into forms which will elude the generalizations which others frame. (MacIntyre 2007:104)

A common thread that runs through all of these different formulations is the assumption that human intentionality is always already contingent and unpredictable even under the most controlled conditions, and that humans are inclined to try to reduce this unpredictability of intentions, especially when others' intentions are at stake.

The creative practices I discuss in this book differ markedly from these scholarly descriptions of ethnographic contexts, in which contingency is used as a cultural resource for negotiating problems of intentionality. The creative practices I describe in this book are informed by modern normative ideals of creativity that celebrate the originality, regeneration, and unpredictability of the intentions of *all agents*. For this reason, the people I worked with consider the contingency and unpredictability not only of their but also of others' intentions as a desirable feature, and they consider the fact that such contingency and unpredictability have become scarce to be a social problem. Theirs are cultural contexts in which overpredictability of others' plans and actions constitutes a problem rather than a desired goal. It is partly because their contexts overflow not only with their own but also with their fellow human beings' overpredictable intentions that they decided to deploy as a solution different digital technologies that can generate contextually meaningful contingency in the field of art.

In itself, the use of contingency as a resource in art production is nothing new. Key figures in aesthetic modernism (such as John Cage in music, Stéphane Mallarmé in poetry, and different visual artists representing Dadaism, surrealism, and abstract expressionism, such as Marcel Duchamp) have used chance procedures in their work. However, scholars have argued that much of this "aesthetics of chance" turns on "the replacement of the desire to do something with the desire to see what will happen" when contingency is integrated into art (Iversen 2010:24). It involves "authorial abnegation" (21), "limiting authorial control" (13), "bypassing of intention" (21), "evading authorial or artistic agency" (12), and "circumventing intentionality" (23; see also Wilf 2013b). John Cage's pioneering work in the 1950s and 1960s, in which he used chance procedures to produce musical and literary work, has been interpreted in this vein. It suggests that there is a strong justification for discussing the design of art-producing computerized systems in relation to mechanical divination and that the connection between the two is not the result of the anthropologist's (i.e., my own) delirious mind:

> Computers provided an excellent vehicle for Cage's work; since his 1953 composition *Music of Changes* he has promoted the concept of nonintention in art, a process in which the artist is no longer required to make decisions in her or his compositions but rather lets chance control creative expression. Initially Cage turned to the *I Ching*, "the ancient Chinese oracle which uses chance operations to obtain the answer to a question," to accomplish tasks for him (Retallack 1996:153). Cage experimented extensively with the aleatoric *I Ching* process, a "discipline" that involved formulating a question and then using coins to divine numbers that provide answers to questions that forced him to ... "break with ego, with habit, with self-indulgence" (Perloff 1991:150). He employed these chance methods as a writer as well, using the *I Ching* to structure poetic lectures and compose poems in the late 1960s. ... [Cage] started using digital technology to extend his practice of making chance-operational texts with computers in 1984. (Funkhouser 2007:64)

Although my interlocutors certainly do take pleasure in the ability of their digital systems to creatively surprise them and in the opportunities such systems give them to problematize their own habitual artistic practices, they do not forgo creative intentionality and authorship. Rather, they harness the contextually meaningful contingency that they can generate by means of digital computation to cultivate creative intentionality and to buttress their ability to become authors. Their intentionality and authorship turn on the ability to engage meaningfully with chance and indeterminacy, in terms both of designing chance or indeterminate events and of knowing how to respond to such events in real time, which does not mean reducing their chance-based nature but rather using it as a creative springboard. Whereas most of the aforementioned anthropological literature has approached the very ability to engage with (mostly in the form of reducing) contingency in its mature state, thereby naturalizing it by leaving out from consideration its conditions of possibility and cultivation in depth, the conditions of possibility for learning to meaningfully design and interact with different forms of contextually meaningful contingency in order to achieve intended and at the same time indeterminate authorial outcomes are at the forefront of my interlocutors' experiments with digital computation.

Against this backdrop, the present book offers an ethnographic analysis of the social construction of creative agency and autonomy, whether of humans or of digital technologies. It asks how, for whom, in what sense, and with what effects digital technologies are designed and understood to be autonomous and creatively agentive, and what answering these questions can tell us about the social construction and cultural embeddedness of human creative agency and autonomy as well. The socially and culturally embedded relationship between control and lack of control, between authorial intentions and unpredictable outcomes, which is a constitutive feature of the use of digital computation for creative purposes, is one of the main concerns of this book.

Another concern is the limitations of art-producing digital systems that are designed to generate surface-level variety and change as indicators and sources of creativity and inspiration at the expense of underlying cultural and social structures and institutions. My interlocutors were first excited by the ability of their systems to generate contextually meaningful contingent output such as notes and words, which they viewed as evidence of creativity and as potential sources of inspiration. However, they soon realized the limitations of a contingent output that is not informed by broader contextual and cultural dimensions beyond, for example, a tune's harmonic structure or a language's grammar. The reasons some of them excluded cultural and societal considerations from their systems are multidetermined. Although my interlocutors readily admitted that it would be extremely difficult to computationally model cultural and social contexts even if they tried to do so, their stance was not based only on technical limitations. First, some of them were motivated to develop their systems because of their dissatisfaction with the social and cultural world. Their efforts thus represented an attempt to escape culture and society. Second, this attempt was compounded by the fact that digital computation has come to represent the perfect platform for generating solutions to social problems because many of its designers and proponents view it as a platform that is itself divorced from and unaffected by the cultural and social world that it seeks to address. This book unpacks some of the unintended consequences of such a belief, especially those that result from the inescapable social embeddedness of the design, use, and evaluation of digital technologies.

Expert Knowledge, Digital Computation, and Technological Solutionism

My interlocutors' efforts to solve the different challenges that they experienced as artists and creative agents by means of digital computation are not merely an instantiation of a general cross-cultural phenomenon of harnessing contingency as a resource for negotiating problems of intentionality. They are also a manifestation of a more restricted cultural context in which technical expertise and technology are approached as offering the solution to every imaginable problem.

Tania Li (2007), drawing on a number of anthropologists who have studied development projects that were orchestrated by experts in colonial settings to improve the living conditions of local populations, has argued that such projects typically involve a number of practices. One practice is that of identifying the problems that need to be addressed and solved. Another, closely related, practice is that of defining and representing the problems and their domains in technical terms in order to make them amenable to expert intervention. The ways in which experts identify and define a problem that calls for intervention anticipate and structure the kinds of intervention they are likely to offer. More specifically, experts are likely to "exclude the structure of political-economic relations from their diagnosis and prescriptions" because they "are trained to frame problems in technical terms" (7). In turn, this exclusion has profound consequences because it "both limits and shapes what improvement becomes" (8). Drawing on Li's insights, Evgeny Morozov (2013) has analyzed Silicon Valley's ideology of technological perfectionism, that is, high-tech entrepreneurs' belief in their ability to improve "just about everything under the sun: politics, citizens, publishing, cooking," by means of digital technologies (5). He has described this belief, which he calls "solutionism," as being based on "recasting all complex social situations either as neatly defined problems with definite, computable solutions or as transparent and self-evident processes that can be easily optimized—if only the right algorithms are in place!" (5; see also Kurzweil 2000; Turner 2006).

The ways in which some of my interlocutors identified the problems they experienced as artists, and the art-producing computerized systems that they designed as solutions to those problems, resonate with these observations about expert intervention and technological solutionism. Thus, although both David and James in their conversations with me contextualized the problems that led them to design their systems in relation to jazz's broader sociopolitical context, which includes the rise of academic jazz education in the United States, in their labs they translated these problems solely into computer-engineering problems. They focused on designing, optimizing, and musically interacting with different kinds of computerized algorithms, thereby devising solutions not in relation to jazz's sociopolitical context and the people who populate it but away from them. In both cases, this approach produced unintended results that intensified a number of dimensions of the problems to which David's and James's systems were meant to provide solutions. Similarly, although some of my interlocutors turned to computer-generated poetry as a way of addressing their displeasure with what they felt to be widespread exclusionary writing practices that are focused on fetishized Romantic notions of creativity, their reliance on digital computation as a creative tool resulted in their reification of another quasi-Romantic creative agency—the computer's—and in their celebration of the rarefied and exclusionary aesthetics that it affords. As the case studies discussed by Li suggest, the exclusion of "political-economic questions" and other messy dimensions of social life (2007:11) is not unique to expert knowledge that is based on the design of digital technologies.[22] However, an expert approach that defines problems and devises solutions solely within the framework of digital computation is likely to accentuate this exclusion because those messy dimensions of social life are not easily amenable to computational representation to begin with.

At the same time, if the efforts made by my interlocutors to solve the different problems they experienced in relation to their and others' creative agency and practice align with these existing descriptions of expert intervention and technological solutionism, they diverge from them in one crucial respect: they represent expert intervention the intended beneficiaries of which are the experts themselves. Hence, whereas anthropologists who have studied the enactment of expertise in different settings have emphasized how experts succeed in absorbing critiques of their knowledge "back into the realm of expertise" (Li 2007:11), "disguising their failures and [continuing] to devise new programs with their authority unchallenged" (10), my interlocutors, by contrast, could not avoid becoming aware of the limitations and failures of their expertise because they experienced them firsthand as "the targets of expert schemes" (11). Against this backdrop, if one of the goals of this book is to describe the conditions "under which expert discourse [and practice] is punctured by a challenge it cannot contain; moments when the targets of expert schemes reveal, in word or deed, their own critical analysis of the problems that confront them" (11), another, closely related, goal is to highlight how efforts that seem to represent the epitome of usurpation of human creative faculties by digital technologies can in fact be the most fragile and open to critique, including critique by the experts who are responsible for such efforts.

The Book's Argument

Based on ethnographic fieldwork in a number of sites over a period of eight years, this book makes a fourfold argument.

First, the people I worked with were motivated to integrate art-producing digital technologies into their creative practices because they wanted to enrich those practices. They felt that such practices ceased to be generative of meaningful newness for different human-centered reasons, both other- and self-focused ones.

Second, my interlocutors thought digital computation had a potential to infuse their creative practices with the meaningful newness that they and others lacked. This potential was based on the mix of control and lack of control that digital computation afforded them. On the one hand, they could control the design of the computational architecture of whatever technologies they used by writing code. On the other hand, they designed these technologies to function as generators of contextually meaningful variety whose results they could not anticipate in advance with complete accuracy. They felt that the variety produced by digital computation, as a kind of nonhuman agent, enabled them to meaningfully rather than randomly extract themselves from what some of them felt to be the impasse of their and others' existing habituated and ossified human-centered creative practices.

Third, however, with time some of my interlocutors experienced and expressed different kinds of frustration with the results of their attempts to replace or enrich human creativity by means of digital computation. Although this frustration found expression in different ways in each ethnographic context, it was generally the result of the trade-offs entailed in the design of art-producing digital technologies. These trade-offs have to do with the prerequisite of abstracting creative practice from many of its context-sensitive and culturally embedded dimensions that make it a form of situated action, as a condition of possibility for its digital representation and simulation.

Fourth and last, these challenges eventually led some of my interlocutors to reevaluate the human-centered creative contexts, which they had tried to enrich and in some cases to escape by means of digital computation, and to seek to reinhabit a number of their dimensions. These dimensions include interacting with human interlocutors, experiencing existing styles that had hitherto been seen as stale, and producing textual specificity and singularity against the backdrop of the abstraction, redundancy, and excess that characterized much of the output of the art-producing digital systems they designed and interacted with. Throughout these contexts, my interlocutors returned from their adventures with computational creativity not only with a renewed appreciation for their previous experiences with nondigital creativity and with a desire to reinhabit it, but with modified and enriched creative capacities and predilections as well.

Thus, the narrative arc that organizes this book points to both the potentialities and the limitations of attempts to integrate digital computation into creative practice in particular, and as an infrastructure of sociality in general. It offers a nuanced critical perspective both on what certain dystopian prophecies describe as the imminent danger of digital technologies' cooptation of and substitution for humanity's creative faculties, and on utopian visions that celebrate digital computation's radically transformational potential for humanity.

Ethnographic Settings, Fieldwork, and Outline of Chapters

Although the practices and normative ideals that inform digital media cannot be abstracted, or only partially so, from the analysis of their underlying technological architecture and their resulting outputs, many critical analyses of new media in general, and of digital art in particular, are based on just such abstract analyses, which leave out the ways in which new media technologies and their products are designed and used in the context of situated action (Funkhouser 2009; Gendolla 2009; Hayles 2009; Hayes 2016a; Hayes 2016b; Miller 2020). For example, although Lev Manovich includes in his definition of "new media" "computer-based artistic *activities*" (2003:13, emphasis added), he argues that the study of "new media" focuses on "new cultural objects enabled by network communication technologies" and "cultural objects and paradigms enabled by all forms of computing" (16). At stake are "cultural objects that rely on digital representation and computer-based delivery" (17) and that can be "reduced to digital data that can be manipulated by software as any other data" (17). These and subsequent definitions and analyses of new media are mostly focused on the formal properties of technologies' architecture and outputs rather than on the real-time situated practices that underlie their design and use. Similarly, Katherine Hayles bases her analysis of "the ways in which distributed cognition is being imagined and instantiated in contemporary electronic literature" on a "tutor text," that is, a specific example of electronic literature (Hayles 2009:39), and Christopher Funkhouser "perform[s] [his] own instinctive and intuitive 'readings' of works" as a way of gaining insight into the aesthetic logic and effects of digital poetry (Funkhouser 2007:6). Although all of these analyses are thought-provoking (indeed, I will occasionally rely on them in subsequent chapters), they are more likely to clarify their authors' interpretative world rather than the ways in which the analyzed technologies and works are perceived and experienced by members of different communities of practice. In other words, they give a partial (albeit important) view of the lived reality of new media, and hence of its meaning. Any technology's meaning depends for its existence, realization, and/or frustration on how individuals use the technology in practice and the ideologies with which they invest it (Gershon 2010). Indeed, as I show in the following chapters, a technology's meaning may even change a number of times for the same individuals during and as a result of their longitudinal engagement with it.

A mostly abstract or formal approach also characterizes the field of computational creativity, or "the study and support, through computational means and methods, of behavior exhibited by natural and artificial systems which would be deemed creative if exhibited by humans" (Wiggins 2019:25). A typical example of this approach's abstract and formal nature concerns the typology of kinds of creativity proposed by Margaret Boden (2003), which has been extremely influential in this field (see, for example, the different chapters in McCormack and d'Inverno 2012 and in Veale and Cardoso 2019), and which I will critically address in some of the subsequent chapters. For example, in relation to Boden's typology, one scholar aims "to give a more uniform and formal treatment of [her] ideas, which can then be used in a precise way," that is, "to give a mechanism through which they can be applied formally (and thence automatically).... The formalisation will ... make it possible to identify desirable and undesirable properties of creative systems in abstract terms. Given the formulation, I define some crucial properties of (artificial) creative systems in terms of the framework, including some which might be proven a priori and some which may be usefully detectable during the activity of the system" (Wiggins 2019:22).[23] As I argue in the next chapters, although it is possible to map the motivations of some of the people I worked with, as well as the intended goals of the digital systems that they designed, alongside Boden's types of creativity (especially in relation to her distinction between creativity that takes place within established styles and creativity that aims to transform those styles), the ethnographic exploration of those motivations and systems as they are designed and used in practice points to the socially constructed nature of those types of creativity, that is, to the fact that they are not abstract properties whose existence or lack thereof can be identified in the operation of this or that system according to formal criteria. Rather, those types of creativity are performatively created and socially constructed in the course and as a result of the longitudinal interaction between, on the one hand, such systems and, on the other hand, their designers and users, who are equipped with historically specific ideologies of creativity and are motivated to design, use, and evaluate such systems in line with those ideologies.[24]

Against this backdrop, this book provides an ethnographically focused analysis of the design and use of art-producing digital technologies as an axis of real-time situated action rather than only of abstract plans and formal properties (Suchman 2007). In so doing, it aims to contribute to the growing anthropological literature that has ethnographically theorized digital technologies and new media (see Coleman 2010 and Gershon 2017 for literature reviews; see also Boellstorff 2008; Coleman 2012; Downey 1998; Helmreich 1998b; Kelty 2008; Kockelman 2016; Madianou and Miller 2013; Seaver 2012).

Beginning in 2011, I conducted ethnographic fieldwork with designers and users of art-producing digital technologies in two ethnographic contexts in the United States. The bulk of this fieldwork, which is the subject of part I of this book, focused on efforts to design jazz-improvising computerized systems in two different academic institutional settings. Chapters 1 and 2 focus on my fieldwork with David, a computer scientist who at the time of my fieldwork worked as a computer science faculty member in a major institute of technology in the United States, where he designed an interactive jazz-improvising computerized system with which he has been playing for many years. Chapter 1 unpacks in detail the compromises that David had to make in using the genetic-algorithms computational framework as his system's basic architecture. It highlights his decision to embed this framework with features of his own style of improvisation instead of using it to generate new and unexpected musical ideas. Chapter 2 focuses on the trade-offs of David's system. On the one hand, David managed to design an interactive system with which he could engage in the kind of tight musical interaction that he felt was missing when he played with human musicians. On the other hand, he ended up with a system that reproduced some of the problems that had motivated him to design it to begin with, especially different dimensions of decontextualized playing. In addition to interviewing David about his research and reading his extensive research publications, I joined him in sessions at his university in which he trained his system by evaluating in real time the improvisations that it generated. Using one of David's trumpets, I also engaged in musical interactions with his system, in the course of which the system and I responded in real time to each other's improvisations on jazz standards. These hands-on playing sessions allowed me to gain a better understanding of the system's capabilities and of David's overall research project. Last, I attended a university function in which David performed with his system in front of hundreds of people. I videotaped and later transcribed all these sessions to account for the multimodal dimensions of David's interaction with his system and for the audience's responses to this interaction.

Chapters 3 and 4 are based on my fieldwork with James, another computer scientist who at the time of my fieldwork headed a research center for music technology in another institute of technology in the United States, where he designed Syrus, the robotic jazz marimba player. Unlike David, James conducted his research with the help of a research team that, at the time of my fieldwork, consisted of two postdoctoral fellows, two doctoral students, six master of arts students, and external contractors who were in charge of manufacturing Syrus's hardware. Chapter 3 focuses on the technical and performative conditions of possibility for Syrus's purported ability to abstract, mediate, and "mix" the improvisation styles of well-known jazz masters. It advances a critique of the notion of stylistic fidelity by showing how the computational abstraction, mediation, and mixing of styles depend on the prior manipulation of exemplars of those styles. This manipulation produces new styles that are more suitable for computational mediation and that bear a distant relation to the preexisting styles that are supposed to be mediated. Chapter 4 develops an anthropological theory of the uncanny to explain how Syrus's embodied form functioned as a new and unexpected source of aesthetic pleasure for team members. Whereas Syrus's embodied form was designed to help tighten the musical interaction between Syrus and its human interlocutors, team members often took advantage of this form to highlight Syrus's interactional incompetence and to undermine the very idea that it could engage in a meaningful interaction to begin with. As part of my fieldwork in this site, I attended James's lab every day for a few months. I was given a workstation next to those of the other team members. I observed different aspects of the research and development and recorded brainstorming sessions. I also conducted semistructured interviews with members of the research team. In addition, I analyzed all the publications that resulted from the team's research, such as journal articles and conference proceedings, as well as coverage in popular media. Last, playing on an electric keyboard hooked up to the computer that animated Syrus's playing, I conducted and video-recorded long playing sessions in which Syrus and I responded in real time to each other's improvisations on jazz standards.

The second part of this book, which encompasses chapters 5–8, is based on ethnographic fieldwork of a narrower scope, which I conducted with practitioners of computer-generated poetry. I focused my participant-observation research on a weeklong workshop that took place in New York City, which I complemented with in-depth interviews with the workshop facilitators and participants, as well as with attending computational literature conferences and forums and analyzing different publications specializing in this art form. The workshop facilitators were key figures in the field of electronic literature in general, and computer-generated poetry in particular. There were fifteen participants, all in their twenties and early thirties. Most had some experience with coding, and all were interested in using digital computation for creative purposes in the field of poetry, although some also experimented with using digital computation for creative purposes in other fields, such as the visual arts.

I chose computer-generated poetry and a workshop dedicated to this art form as my second ethnographic focus for a number of reasons. First, writers of computer-generated poetry must learn to engage with digital computation by writing code. In contrast, a jazz musician's playing when he or she interacts with David's system or with Syrus is not completely or fundamentally different from his or her playing when he or she interacts with human musicians or plays by himself or herself (although, as I will clarify, there are significant differences). The two ethnographic contexts thus provide an opportunity to explore distinct topics. In the case of jazz, issues such as mimicry, originality, and sociality, which are related to the real-time musical interaction between a human musician and a jazz-improvising digital system, take center stage, whereas the case of computer-generated poetry raises questions around creative intentionality in relation to the affordances and limitations of computational indeterminacy and potentiality as they find expression in the practice of writing code whose execution by the computer results in output texts the exact nature of which the poet cannot know in advance with complete accuracy.

Second, a workshop that focuses on computer-generated poetry functions as a site in which neophyte poets are socialized into writing this form of poetry as a distinct form of creative practice by means of explicit metapragmatic instructions, hands-on experimentation, and critique of the results of this experimentation. It provides a good vantage point from which to document some of the normative ideals and practices that inform computer-generated poetry, including whether these practices are aligned or misaligned with these ideals. In addition, because many of the workshop participants are also writers of nondigital poetry, the workshop provided a window into their experience of the differences between writing nondigital poetry and writing computer-generated poetry, and thus into the distinguishing features of both forms of writing. During the workshop I engaged in all the same activities as the other participants, including writing and critiquing computer-generated poetry and attending social events outside of the workshop. I informed my analysis by ethnopoetics as a form of study that "should engage in a principled dialogue with local theories of meaning and moral responsibility, local interpretative frameworks [and that] recognizes narrators as coeval, fully intersubjective in the doing of things with words" (Webster 2020:87). In this framework, poets are understood as "both creative individuals and socialized language users imbricated in wider social, political, cultural, and historical fields" (Webster 2015).

Rather than providing an exhaustive analysis of the field of computer-generated poetry, one of the main goals of the second part of this book is to present a comparative perspective against the backdrop of the insights I raise in the first part that focuses on jazz-improvising digital systems. In chapter 5 I describe some of the key aesthetic and technical dimensions of computer-generated poetry that provide the backdrop for how workshop participants experienced the potentialities and limitations of this art form, which I explore in chapters 6–8. Chapter 6 focuses on the ways in which some of those potentialities and limitations are informed by the tensions of American liberal subjectivity, whereas chapters 7 and 8 contextualize them against the backdrop of the aesthetic principles and creative practices of the French literary group Oulipo, which some of my interlocutors take as their immediate inspiration. By means of a close analysis of my interlocutors' practices and their comparison with the ideas and literary products formulated and created by members of the Oulipo in the 1960s, I show that the cultural contradictions and tensions that accompany attempts to simulate and enrich creative behavior by digital computational means are not entirely the result of, and hence do not entirely depend on, the realization of those attempts by means of digital computational technologies although such a realization can intensify those tensions and contradictions. Rather, they already inhere to some degree in the general orientation to creativity as something that can be simulated or enriched by abstract, formulaic, mathematical, and algorithmic means.

Another comparative dimension provided by the analysis of computer-generated poetry has to do with one of the reasons my interlocutors are drawn to the computer as a creative partner, namely, the fact that it can function as an easily available source of pure contingency that can be used to generate variety and change, which can then be interpreted and perceived as indicators and sources of creativity and inspiration. A significant limitation of this approach is that it might result in surface-level variety and change at the expense of the kinds of underlying cultural and social structures and institutions, as well as other sources of contingency, which give creative behavior its meaning for its practitioners. In my fieldwork experience, the limitations of relying predominantly on the generation of surface-level variety (limitations often experienced as the problem of meaninglessness) became much more salient for the practitioners of computer-generated poetry than for the practitioners of computer-generated jazz. Some of my interlocutors in the field of computer-generated poetry perceived the use of pure contingency for creative purposes as a problem because they tended to compare its literary products with the products of non-computer-based genres of poetry, in which each word is typically the result of a careful choice and deliberation, whereas my interlocutors in the field of computer-generated jazz were more tolerant (up to a point) of the occasionally meaningless results produced by this kind of contingency, because contingency is the very stuff of which jazz, an art form based on real-time group improvisation, is made (Berliner 1994:210–216). By focusing on two ethnographic contexts that are different from each other in different ways, I show the contextual specificity alongside the shared characteristics of the potentialities and limitations of designing and using different creative computational systems.

Although artists design and use digital technologies in many creative fields that can be compared to one another, I designed my fieldwork around these specific ethnographic contexts of computational creativity for three additional reasons. First, each of the two jazz-improvising systems that I focus on in this book occupies an important place in the history of the development of generative (and robotic) music systems and has received extensive media coverage. They are thus likely to provide a good perspective on the cultural complexity that informs attempts to develop art-producing computational technologies, although by now there are more advanced technologies designed for similar purposes.[25] Second, historically, computer-generated poetry marked the beginning of digital literature at large. Inasmuch as the workshop facilitators chose to organize the workshop around a number of classic computer-generated poems that were written in the late 1950s and 1960s, the workshop provides a window into aesthetic principles and sensibilities that have continued to inform not only the cultural order of digital poetry but also, to some extent, of digital literature in general (Funkhouser 2007:6–7, 41). Last, my choice to conduct ethnographic fieldwork in these two contexts has been motivated and informed by my three previous ethnographic projects on the institutional transformations of creativity. These projects include an ethnographic fieldwork on nondigital creative-writing workshops (Wilf 2011; Wilf 2013c), on the institutionalization of jazz music in academic jazz music programs (Wilf 2014a), and on the work of business-innovation consultants (Wilf 2019). It was also informed by my concrete experience as both a semi-professional jazz trumpet player and as a published poet, which has given me a better understanding of the intricacies and stakes of the design and use of digital computation in jazz and poetry.

However, it should be clear from the outset that my insights into these intricacies and stakes are not meant to be—and, indeed, cannot be—exhaustive, for three main reasons. First, the field of art-producing digital technologies is too vast for any book to cover in depth, especially a book that provides a comparative perspective on different ethnographic sites focused on two distinct art forms. Second, this comparative perspective necessitates the kind of relatively short ethnographic fieldwork excursions that have become common in contemporary multisited ethnographies of late Western modernity and that cannot be as exhaustive as ethnographic fieldwork that focuses on a single ethnographic site and community of practice. Last, this book addresses a field that is constantly changing owing to the emergence of new technologies. Significant technological changes can transform this field from the moment an author has finished writing a book to the time the book sees the light of day.

These factors notwithstanding, one of the key arguments I make in this book is that although human beings may design ever new and different art-producing or other forms of digital technologies, these technologies will remain rooted in specific social predicaments that will continue to haunt them by informing the motivations for their design, their architecture, humans' experience of interacting with them, and humans' evaluation of their output in relation to the problems that they were meant to solve. Hence, rather than trying to provide a description of the latest and most sophisticated art-producing digital technologies—a description that is likely to become obsolete pretty quickly anyway—this book focuses on the enduring dimensions of the development, use, and evaluation of such technologies, which result from their irreducible and inescapable embeddedness in culture and society. Against the backdrop of the fact that our contemporary moment is informed by the increasingly prevalent idea that the design of better and more sophisticated computational technologies holds the key to a better—that is, more "efficient and rational"—future, compared with a socially "messy" present, this book points to a different conclusion. It shows that computational solutions to real or imagined problems and insufficiencies are best developed in relation to the specific social and cultural contexts that gave rise to those problems and insufficiencies, rather than away from them.

PART I

Jazz
Mimicry, Originality, Sociality

1

"I Prefer Playing with It to Playing with Most People":
The Computer as a Musical Conversation Partner

OD'ing on Clifford Brown

I was sitting in an Indian restaurant with David, a computer scientist in his late fifties. David, who at the time of my fieldwork worked as a faculty member in an institute of technology in the United States, had over a span of many years developed a jazz-improvising digital interactive system. The code that he wrote runs on a portable Macintosh from the early 1990s, and its resulting improvisations are played by a simple tone generator. After we ordered, I asked David to explain what had led him to develop this system. He responded that its origins date back to a research project he had initiated out of pure curiosity during a sabbatical. Over time, however, he realized that what had initially started as a theoretical interest might become a practical solution to a problem that had for long frustrated him as a semiprofessional jazz trumpet player. He explained to me in an animated voice:

> I prefer playing with it [the system he built] to playing with most people because I get a bigger kick out of how it responds to what I play than out of what people do to what I play when we trade four-bar phrases. And I get much better ideas listening to it formulating a response than I do by listening to most people, because most people play licks [musical phrases] that they know instead of making something up [in response to what I play]. Jazz is supposed to be about making something up. I remember going to jam sessions and there's often no [musical] conversation [between players]. Everyone is just playing the licks they learned. You'd get a trumpet player who has clearly OD'd on Clifford Brown and this is all he plays.[1] And he's [Brown] a good trumpet player to OD on, but still! I'd like to hear something other than Clifford Brown. And I could listen to him [the player] and say: "Oh, god, yeah, he's going to do this thing [lick] now, there it is. Yeah, he just did it," you know? So I get more inspiration from playing with my system than with people.

David became more personally invested in his system against the backdrop of his growing disappointment with the jazz musicians with whom he had been playing. His disappointment turned on his feeling that they confined themselves to playing the same musical vocabulary that they had learned instead of opening themselves up to the musical ideas that their human interlocutors played in the real time of their joint improvisation and instead of engaging in a meaningful musical conversation with them according to the normative ideals that are supposed to inform the jam session as a form of creative activity in the jazz cultural order (Monson 1996:84).

The atomistic kind of playing that was at the core of David's disappointment is the result of concrete historical changes in the modes of jazz training in the United States. The dwindling commercial demand for jazz since the 1950s has resulted in the gradual disappearance of performance venues in which neophyte musicians could cultivate their skills by means of listening to and apprenticing with experienced practitioners in live performance settings. This apprenticeship-based form of training has gradually given way to abstract, formalized, and rationalized education in the many academic jazz programs that have replaced the commercial jazz scenes of the past and that have become the institutional pillars of the contemporary jazz world and in which many jazz musicians learn their craft and hope to find some form of employment (Chevigny 2005 [1991]:51–52; Rosenthal 1992:170–173; Wilf 2010:572; Wilf 2012:35–36). These institutional transformations—the shift of jazz training from the clubs to the classroom, where standardized, printed pedagogical aids have become students' key mode of acquiring knowledge—has led to the neglect of the pragmatics of real-time musical interaction between players that requires intensive listening to one another (Wilf 2014a:167–169, 177–181, 202–209). The result has been the perception, shared by academic jazz programs' administrators, teachers, and students, that many young musicians are content with playing the same standardized licks that they have mastered, and that they do so in a very formulaic, solipsistic, and decontextualized way that runs against the aesthetic core of jazz as an improvised form of music in which each player must learn to inform his or her improvisation by the playing of his or her bandmates.

David was frustrated by this lack of musical conversation and interaction between players. Consequently, he set out to create a jazz-improvising digital system that could tightly "listen" and respond to his playing, and to whose playing he could tightly listen and respond—all in the real time of improvisation. On the surface, his goal resonates with one of the origin sites of artificial intelligence, namely Joseph Weizenbaum's 1960s research on the possibility of designing the computer to engage in a natural-language conversation with humans (Weizenbaum 1966). Weizenbaum programmed the computer to respond to the queries of the human user as if he or she were a patient and the computer a Rogerian therapist. However, Weizenbaum's system generated a lot of scholarly attention because its human users became invested in the interaction with it, despite the fact that its responses were more often than not repetitive, ambiguous, or simply nonsensical.[2] In contrast, David designed his system based on a complex computational framework that ensured that the system's musical responses to David's playing would always be unexpected and yet always make musical sense. This computational framework, as well as the fact that the conversation that David wanted to orchestrate takes place in the much simpler domain of music rather than in the domain of natural language, were key dimensions of his system's success. At the same time, they also created new complications that he had not anticipated in advance.

Building a "Genetic" Code of Improvisation

David chose to design his system predominantly by means of the genetic-algorithms computational framework. This framework borrows its conceptual machinery from the domain of evolution and natural selection, particularly with respect to animals that reproduce sexually (Helmreich 1998a). It is usually used for optimization, that is, for finding the best solutions to specific problems (McCormack 2012). First, an initial population of potential solutions to the problem one is attempting to solve is generated at random. Each solution is represented as a string of bits, which functions as a kind of "chromosome." This group of solutions is known as "generation zero." Each of these solutions is then assessed in terms of its ability to solve the problem and assigned a fitness level. Solutions with a higher fitness level are chosen to function as "parents" and "mate" with each other. In the process of mating, the two parents recombine parts of their chromosomes with each other and thus produce two "kids" that replace solutions with a lower fitness level. This produces subsequent "generations" of solutions, with the "population" size remaining constant. The programmer might choose to introduce "mutations" in the chromosome structure of some members of the population by randomly switching some bits from zeros to ones and vice versa. The process continues for as many generations as needed until the generated solutions converge around a specific solution and/or the programmer is satisfied with the population of solutions, at which point he or she terminates the process.

One of the main reasons David chose this framework to design his jazz-improvising system is that the conceptual basis of this framework naturally lends itself, at least on the surface, to modeling how mature jazz musicians think about improvisation. Jazz is an improvised form of composition; the improvisation occurs on the spur of the moment, and in the course of it a musician weaves together preexisting and conventional, or genred, musical building blocks such as short phrases and motifs (the "licks" that David mentioned time and again in our conversation) in accordance with a given harmonic progression and in response to the real-time contribution of the musician's bandmates. This compositional logic is shared by many forms of improvisation in different ethnographic contexts and historical periods (Finnegan 1998; Lord 1960; Sawyer 1996). Every musician is expected to acquire and master this reservoir or "vocabulary" of musical fragments, as well as the ability to creatively use them in different harmonic contexts (Berliner 1994; Monson 1996).[3] To do so, neophyte musicians are encouraged to listen attentively to the recorded improvisations of the past masters, to identify the building blocks of those improvisations, and to practice them in different harmonic situations and keys and thus incorporate them into their own playing (Wilf 2010:564). The fact that a key dimension of jazz improvisation consists of the creative mixing and recombination of a stock of short, conventionalized musical phrases or licks—each lick ranging from a few beats to a few measures in length—encourages a metaphorical slippage that naturalizes thinking about such licks as "genes," whose recombination and mutation can result in more musically satisfying licks that, when properly used with respect to a harmonic progression and the playing of one's bandmates, can contribute to more satisfying or at least new improvisations.[4]

Accordingly, David defined two types of coevolving populations, a population of phrases and a population of measures. In the phrase population, which, for the sake of simplicity, will be my primary focus, each individual "solution" or member of the population is a musical phrase, or lick, four measures long (in 4/4); the population size is forty-eight phrases. Following a conventional body of jazz theoretical knowledge known as chord-scale theory, David assigned to the different kinds of chords that might appear in most standard tunes specific scales that correspond to them. The program he wrote can read the harmonic progression of a tune and adapt the phrases to the different chords that this progression consists of by mapping the phrases—which are basically musical contours—to the scale assigned to each chord.

In the initial version of his system, the phrase population was developed in the following way. First, an initial population of forty-eight phrases was initialized by generating random bit strings of the appropriate length for each phrase. The system then improvised a number of solos, each solo consisting of three choruses of a thirty-two-bar form, which equals twenty-four phrases, that is, half of the phrase population. During these solos phrases were selected at random. David himself evaluated the fitness of the phrases (and, simultaneously, of the measures) by sitting in front of the computer and hitting the B key (the B stands for "Bad") to lower the fitness level of a phrase/measure, and the G key ("Good") to raise it, at the same time that the system improvised. In a subsequent "breeding" mode, after the population was sufficiently sampled and initial fitness levels were assigned to the different members of the phrase population, four individuals were chosen at random regardless of their fitness to form a "family." The two individuals with the highest fitness levels in each family were chosen to function as parents, whose two "offspring"—generated by means of recombination at a random crossover point along each phrase's bit string—replaced the two family members with the lower fitness levels, resulting in the replacement of half of the population. The system then improvised three more choruses, and David assigned fitness levels to the phrases. A new generation was then produced by means of the same process of breeding according to the new fitness levels and so on until a phrase population that David deemed good enough (which he calls "a soloist") was produced.

Being Creative with Genetic Algorithms in Every Possible Way

In using genetic algorithms to enhance his creativity in jazz, David had to get creative about enhancing their capacity to perform the task he was assigning them. That is, he had to deviate from the standard genetic-algorithms framework. The reason is that, to reiterate, genetic algorithms are best used for optimization, that is, for finding the best possible solutions to specific problems in domains such as engineering, in which it is assumed that there is an optimal solution that can be assessed objectively given specific problems, and in which it is expected that the evolving generations of solutions will gradually converge around this optimal solution (Ritchie 2019:181).

However, when applied in the domain of creativity and aesthetics, this framework presents two problems (McCormack 2012:39). First, modern Western normative ideals of creativity in general, and normative ideals of creativity in jazz in particular, emphasize the importance of diversity, newness, and uniqueness on a number of levels, rather than convergence around a single or a few solutions. For example, each jazz player is expected to have a unique improvisation style, which means that when two players improvise on the same harmonic progression they are expected to produce different improvisations (Wilf 2014a:170–174). Furthermore, such diversity should exist and be displayed not only across the improvisations of different players, but also across the different improvisations produced by a single player, and, finally, even during a player's single improvisation. Although having a unique and distinct improvisation style implies a certain level of redundancy and repetition—for this is what makes a style recognizable, to begin with—players are not supposed to generate improvisations that are too repetitive and clichéd. Indeed, although David was led to develop his system primarily because the musicians with whom he had played did not engage in a meaningful musical exchange with one another, he was also bothered by the related fact that the improvisations they produced were similar to one another (because they were derivative of the playing of the same past jazz masters) and repetitive on the level of the single solo.

David told me that to address this problem and to maintain diversity of phrase populations, he had to put a lot of thought in the design of the mutation operators:

> There's a lot of stuff going on to ensure that it [his system] is not going to play the same thing over and over again. For example, if a given measure gets played once, it gets mutated before that measure might ever get played again. Not to mention that in the breeding version, the mutation operators ensure diversity, and I really stood on my head there. Because you really have to go against the grain to defeat the schema theorem.[5] Because the schema theorem is really powerful. It wants to converge, it really does! And you can't have that because then you'd have the same minor variations and that would be a disaster in an aesthetic context.

David made sure that the mutation operators, which in the context of the standard genetic-algorithms framework are supposed to be random processes (e.g., randomly switching some bits in a chromosome from zero to one or vice versa), would be "intelligent," that is, that their logic would be aesthetically motivated. He designed two mutation operators specifically to counteract the convergence problem and to ensure diversity of the musical phrases. One mutation operator substitutes a randomly chosen measure for the measure in the phrase that occurs most frequently in the phrase population, thus making sure to reduce the occurrence of overly successful measures. Another mutation operator generates new phrases from the least frequently occurring measures in the phrase population, thus making sure that diversity is maintained.

The second problem in using the genetic-algorithms framework in the domain of creativity and aesthetics is that, as opposed to engineering contexts, where it is possible to assess whether a potential solution is optimal or not given a specific problem and by means of objective standards and criteria, excellence in the field of aesthetics and creativity is notoriously difficult to define, evaluate, and measure in any objective way, in part because there are no clear problems against which solutions can be assessed (Wiggins 2019:35; Wilf 2013c:136).[6] Consider the following description of the lack of clear criteria of excellence in the field of poetry: "It is true that strict standards of competence are applied by literary critics, but even here the criteria are amorphous. Writing is a sedentary activity; the poet grows no calluses, goes to no offices, punches no time clock, gains no diploma, earns no certificate of competence. *There is no current universal measure of success or fitness*; credentials are intangible, a passport without date, country or occupation. Poetry is an undefined profession. It lacks both the institutional support and the popular appeal of science" (Wilson 1986:68, emphasis added). The use of "fitness" in this description is serendipitous. Not only poetry but jazz, too, lacks a "universal measure of ... fitness" that can be programmed to assign fitness levels rigorously and consistently to musical phrases and measures.[7]

To solve this second problem, David did two things. First, some of the other mutation operators that he designed are, in his words, "musically meaningful," in that they consist of compositional techniques such as rotation, transposition, retrograde inversion, and retrograde reversal, which have long been used in different musical genres such as Western classical music, jazz, and serialism. The purpose of these operators is to "quicken the learning process" by creating "not just new, but better offspring." In effect, David embedded his own specific aesthetic criteria and preferences in the mutations themselves. In doing so, he determined the nature or essence of the resulting phrase population in advance, and in violation of the genetic-algorithms framework. As we shall see below, during the design of subsequent versions of his system, David went even further with embedding his own specific aesthetic criteria in the computational process. Second, as I noted above, rather than tackle the problem of programming a fitness function, which he deemed insurmountable, David chose to assign fitness levels himself to the phrases (and measures) that his system generated and executed. During the system's improvisations, he would hit B to lower the fitness level of a phrase/measure and G to raise it. In this case, too, he shaped the emergence of the phrase population in a way that strongly determined the nature and essence of this population in advance.

"This Is a Dream Come True for Me":
First Impressions of Being a Divine Aesthete

The first version of David's jazz-improvising digital system provided him the opportunity to realize his dream of having total control in jam session–like situations. Immediately after telling me about a typical jam session where he would encounter a trumpet player who "has clearly OD'ed on Clifford Brown," David said: "But with my system, if I hear something that's annoying, I can just train it out—bad, bad, bad, bad, bad—and then breed it out, which is nice." As he was saying "bad, bad, bad, bad, bad," David made the gesture of hitting a key on a computer keyboard with his finger to simulate the action of providing fitness levels to his system's phrases by hitting the B key. In an interview in a radio show a few years before my fieldwork with him, David had commented: "This is a dream come true for me, because I've played so many jam sessions where you'll get a guy who comes in every week and plays the same licks and you'd like to say 'bad bad bad bad' and breed it out of him. And I can do it here [with my system]. So if there's an annoying lick—boom, it's out of here. It's fabulous!"

David's pleasure in the control that his system afforded him resonates with critical studies of computer-based "artificial life"—that is, algorithmic processes that model life processes—and of the scientists who design them. Many programmers are attracted to programming microworlds because "in the microworld, the power of the programmer is absolute" (Helmreich 1998b:69). Stefan Helmreich interviewed a programmer who "reflected self-consciously that he built simulations because he wanted to get away from a world that felt difficult to control" (70). He argues that one cultural context that anchors this and similar statements is "a normative kind of Euro-American masculinity that consists of emotional disengagement, escape and autonomy from others, calculative rationality, and a penchant for objectifying, instrumentalizing, and dominating the world" (ibid.). Such a context runs parallel to other contexts such as "Western-Judeo-Christian cosmology" (83), which provides programmers the conceptual machinery to think of themselves as "a genus of god"—creators and destroyers of worlds (84). Such a god "betrays a desire for the scientist to become an omnipotent and omniscient god, to make this a practical reality in an artificial world that can be perfectly monitored and controlled" (85).

Although David never used the notion of a god to describe himself to me in relation to his system, the kind of control that his system gave him and that he relished resonates with the god imagery and its function as described by Helmreich. By taking it upon himself to assign fitness levels to different members of the phrase and measure populations—that is, deciding whether they "live" or "die"—David functioned as an even more omnipotent kind of god than the programmers described by Helmreich, who tend to limit themselves to defining the initial algorithmic processes that provide the basis for the subsequently emerging simulated "life." In the microworld that David designed, he functioned as a god not only on this general level, but also on the level

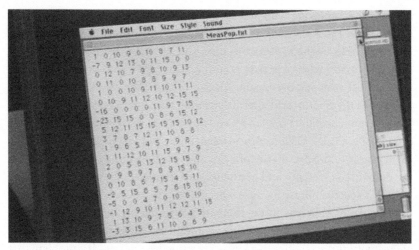

FIGURE 1.1. A file displaying the results of David's evaluation of a measure population.

of the specific members of the phrase and measure populations, a god who micromanages life and intervenes in it to obtain a much higher resolution of control in every stage of the process of its emergence. For example, after he demonstrated to me how he used to develop phrase and measure populations by listening to the improvisations generated by the early version of his system and hitting the B and G keys according to whether he liked or disliked what he heard, David opened a file titled "MeasPop," which displayed data on the resulting measure population (see fig. 1.1). While looking at the data, he noted: "Here's one that's really not a good measure. It got minus 23 [David refers to the first numerical figure that appears on line 8]. So that one got twenty-three more bads than goods. That's pretty bad. This guy is not going to make it to the next round." As David was speaking, he burst out laughing. Thus, within the confines of the specific microworld that he programmed, David has come to function as a kind of divine aesthete who has the power to determine the fate of specific phrases and measures.[8]

Helmreich adds that "god imagery also inhabits descriptions of postcreation events.... Interventions after creation are routinely dubbed 'miracles'" and "divine intervention" (1998b:85; see also Wiggins 2019:36). Although, to reiterate, David never explicitly referred to himself as a god vis-à-vis the microworld that he had programmed, he too performed "postcreation events." Thus, a few minutes after talking about "the guy" (i.e., measure) that "is not going to make it to the next round," he told me: "I'll look out for licks that are really recognizable. And what I did the other day, I had a soloist [specific phrase and measure populations that were the result of previous training sessions] I was playing with and I kept hearing the same lick and it just got on my nerves so I just took it out of the database. I don't use that measure any more. [He laughed.] Which is nice, you can have that luxury. You don't have that luxury when a person is sitting in [in a jam session] and they play the same lick over and over again." David's postcreation event, that is, making changes in the microworld after its creation, takes place on the level of individual members of the population, and not only on the level of the basic "conditions in a model" (Helmreich 1998b:85). The purpose of this postcreation event points to the peculiarity of using the genetic-algorithms framework in the domain of creativity and aesthetics, as I discussed above. David decided to "kill" a specific phrase because it became too successful, that is, it occurred too frequently. Whereas in the domain of engineering, for example, such a success would typically indicate that a solution has been found, in the domain of creativity and aesthetics this kind of solution becomes itself a problem that needs to be addressed.

"It Was Pretty Deadly": Second Impressions of Being a Divine Aesthete

As opposed to the computer scientists who program microworlds over which they have control but that are markedly different from the real world in which they live, David created a microworld in the image of one subset of the real world that he himself used to inhabit: the world of jam sessions. His purpose was to have in this microworld the kind of control that he lacked in the real world, specifically the ability to eliminate "annoying licks" played by his musical partners. The result was "a dream come true," he told me.

However, David soon realized that he had to pay a heavy price for his dream of exercising complete control over his musical environment. Although the dream of being able, godlike, to "breed off annoying licks" with the mere stroke of a B key initially enticed him, once realized it soon turned out to be a tedious nightmare. When I asked him to describe the experience of training a soloist in the early version of his system, he responded:

> In the early days it would have lasted a couple of hours. It was pretty deadly. What I found early on was that if I use the simple uniform random generator for training purposes, then the average interval [the interval between two adjacent notes in the phrase population] is about a seventh. It [the typical phrase] just hops all over. And that's pretty much an octave. And the average note distribution would be flat too. There would be as many high notes as middle notes as low notes. Very chaotic. It sounds really chaotic because it is generated randomly! The notes are in tune, by definition, but they don't sound particularly melodic to my ears. So when you use this simple uniform

random generator population, it will take three or four generations before you get something that sounds at all musical. So you're sitting there and go "bad bad bad bad bad bad bad bad. Oh, OK, I'll lay off a little bit. That wasn't too horrible. Oh, that was kind of good. Good good. Bad bad bad bad bad bad," you know, and it's a really low yield.

When David began to evaluate the fitness of the licks produced by the system, fantasies of complete control gave way to numbness and exhaustion. Because the early version of his system produced the generation zero of phrases randomly, these phrases were utterly chaotic and lacked any kind of musical sense. Assigning fitness to them was akin to listening to white noise and trying to assess the aesthetic value of each of its different fragments relative to all of its other fragments—an impossible task. Compared to such random phrases, the "annoying licks" produced by the human musicians whom David had tried to escape by building his system were pinnacles of aesthetic sophistication. In the interactive genetic-algorithms framework, David's predicament has come to be known as "a 'fitness evaluation bottleneck' that reduces the human operator's role to that of a 'pigeon breeder' who quickly fatigues" (McCormack 2019:333). "Levels of expertise, fatigue during system runtime, individual bias in preferences and varying levels of concentration" are some of the common problems with human-based fitness assignment (Jordanous 2010:226).

The limits on his cognitive ability to listen to the randomly generated phrases produced by his system and to evaluate their fitness motivated David to design his microworld with these limits in mind. He explained:

> When I do the training thing, it will replace half of the population in the breeding mode for each generation. After it does this, you need to listen to it so you can provide fitness. So what it does is, it plays the tune using initially the newbies, the ones that were just created. It plays them back because you have to hear them now. And if the tune is still going on and I run out of those then I'll be using others too and provide fitness for those. So if the phrase population is forty-eight phrases, that means half of the population is twenty-four phrases, twenty-four four-bar phrases. Well, if it's a thirty-two-bar popular song form, that's three choruses. And that was all right then because I had trouble paying that close attention for more than three choruses. After four or five choruses of those phrases I would zone out—I can't concentrate that long listening to those licks! [As he was saying this, David raised his voice in exasperation.] So that's where the forty-eight phrases came from, the phrase population, because I could only stand those for three choruses in the early generations.

David reveals that the size of the phrase population—forty-eight phrases—was not the product of an arbitrary decision, but rather the outcome of how long he could concentrate while listening to the highly unmusical licks created by the system in the first few generations. To push the metaphor of the programmer as god a bit further: David's creation became a reflection of his own human limitations rather than a site in which he could feel like an omnipotent god who knows no limits.

Furthermore, even just to get to a state where the system could generate a reasonably musical generation after "three or four generations," David had to deviate from the genetic-algorithms framework in some of the ways I mentioned above:

> So you figure three choruses of a thirty-two-bar standard tune, let's say in a tempo of 150, it's, what, four minutes, less? And then you have to do the next generation. And you take a break in between. And maybe you go out and cheat a little bit, you go out and mess around with the data file. So you can cheat and train a soloist in an hour, easily. Because otherwise it's painful. So it's not that bad, but that's because the mutations are so good. What's going on is that the mutations are really doing good things. And that's key because without the intelligent mutations they would act like regular genetic algorithms, they would have hundreds of generations of huge populations just to get something in the ball park. I don't have time for that. They better sound good NOW or I'm out of here! I don't want to—I can't!—listen to crap for that long. And that's initially what led me toward those mutation operators.

A number of things are worth noting about David's explanation. First, to reiterate, rather than have mutations that operate randomly as is expected in the genetic-algorithms framework, David programmed "intelligent" mutations, that is, mutations that are aesthetically motivated, in that they are based on compositional techniques that have been used in a number of musical genres. The reason David deviated from the genetic-algorithms framework, it turns out, is the numbness that he felt when he had to assign fitness levels to the chaotic licks produced by his system.

Second, when David talks about changing the program in mid-action, that is, after he programmed his microworld and set it in motion, he does not use a terminology that is reminiscent of the notions of "miracles" and "divine intervention" that programmers typically use to describe their "postcreation events" (Helmreich 1998b:85). Rather, David uses the notion of "cheating" to describe the process of "taking a break" from training a soloist and "messing around with the data file" in order to shorten the time it takes to reach a reasonably musical population of phrases and thus to make the process less numbing. As opposed to the notions of miracles and divine intervention, which support the idea of an all-powerful god who is in complete control of his or her creation, the idea of cheating suggests an agent who lacks control over his or her creation and has to scramble around in an ad hoc and messy way because this creation does not yield the results that he or she had hoped it would, and because he or she cannot cope with the resulting disorderly microworld.

Third, on the surface, David's impatience with the time it would take his system to generate or "evolve" reasonably musical phrase and measure populations and his desire to precipitate this evolution's pace resonate with the fact that a key reason many programmers develop microworlds that are modeled after evolutionary processes is their impatience with the slow pace of evolution in the real world. For example, Helmreich discusses the factors that led Tom Ray to develop Tierra, a highly popular microworld that draws from the conceptual machinery of evolution by way of genetic algorithms: "Ray holds that Tierra is an alternative biological world and is a satisfying site to examine evolutionary processes. Originally trained as a tropical biologist at Harvard in the 1970s . . . , Ray has done extensive work in Costa Rican rain forests but has been frustrated by the pace of evolution there, which happens too slowly to reveal to him the ecological dynamics in which he is interested. Ray sees Tierra as perfect for doing experimental evolutionary biology" (Helmreich 1998b:112). Similarly, consider the way Ray Kurzweil, an influential techno-futurist, describes evolution in the biological world and compares it to its computer-based simulation:

> Evolution has achieved an extraordinary record of design, yet has taken an extraordinary long period of time to do so. If we factor its achievements by its ponderous pace, I believe we need to conclude that its intelligence quotient is only infinitesimally greater than zero. An IQ of only slightly greater than zero (defining truly arbitrary behavior as zero) is enough for evolution to beat entropy and create wonderful designs, given enough time. . . . Evolution is thereby only a quantum smarter than completely unintelligent behavior. The reason that our human-created evolutionary algorithms are effective is that we speed up time a million- or billionfold, so as to concentrate and focus its otherwise diffuse power. (Kurzweil 2000:44)

Kurzweil's broader thesis is that technology is "the evolution of (biological) evolution" (Kurzweil 2000:43). Technological (especially computer-based) evolution is a much quicker and more efficient form of evolution than biological evolution, and it represents a new stage in the human species' own evolution.

Against the backdrop of such descriptions of technologists' impatience with the slow pace of biological evolution and their celebration of computer-based evolution, David's experience with his system offers a different and telling perspective. David became frustrated not with the slow pace of biological evolution, but with the slow pace of the computer-based simulation of evolution that he had designed and programmed. In his case, it was computer-based evolution rather than biological evolution that failed to produce satisfying results in a timely manner. If Tom Ray and other programmers hope to solve the problem of evolution's slow pace by shifting from the world of biology to that of technology, David's shift from the world of biology to that of technology was a key determinant of this problem. Finding a solution required him to shift back from the world of technology to the world of biology, as it were—that is, to intervene in the computer-based evolutionary process as a representative of the biological world by "cheating" and "messing around with the data file" and thus "help" the technological simulation of evolution to quicken its pace.

To reiterate, one reason for these complications is that David took it upon himself to assign fitness levels to members of the phrase and measure populations. Because earlier generations of the populations were so chaotic and unmusical to his ears, the task of assigning fitness levels proved unbearable, and he experienced the pace of the evolution of the populations as being too slow.

However, as became clear when I joined David for a few training sessions of phrase populations generated by the version of his system that already integrated the intelligent mutations he had designed to quicken the pace of evolution, the task of providing fitness levels remained a daunting task, for two reasons. First, I noted above that the idea that jazz improvisation consists of the creative concatenation of short, conventionalized musical phrases or licks—each lick ranging from a few beats to a few measures in length—encourages a metaphorical slippage in the context of which it is possible to think of such licks as genes, whose recombination and mutation can result in more musically satisfying licks. In other words, it motivates the turn to the genetic-algorithms framework as an appropriate computational architecture for computationally simulating jazz improvisation. However, one of the main problems with this approach is that it results in a musical output characterized by what I call "low-level meaning," that is, musical meaning that exists mostly on the level of half a measure or one chord—a point I will elaborate on in detail later in this chapter—and consequently whose evaluation by a human listener becomes a very frustrating task. This task is frustrating because such short musical fragments have very little aesthetic meaning to human ears. It is only by competently concatenating such fragments into larger units and in view of higher-level compositional considerations that musical structures can be created, for which human ears can have a preference.

During our joint training sessions, after David trained a soloist for a few choruses, I replaced him and continued to train the same soloist. As I listened

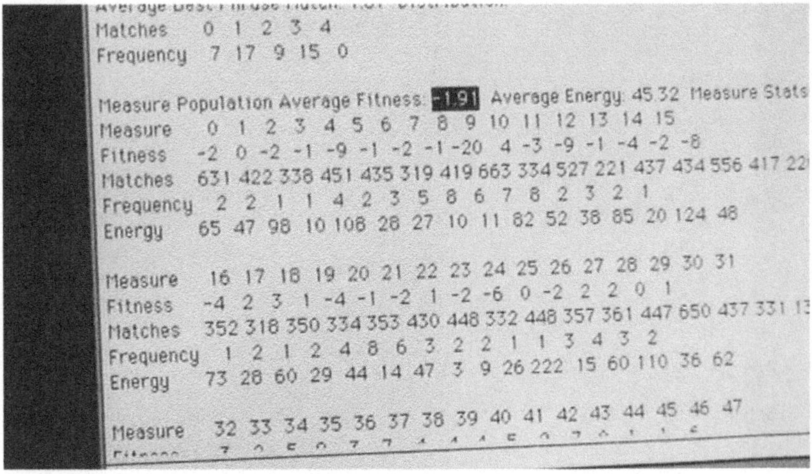

FIGURE 1.2. A file displaying different statistical data on a trained measure population.

to the system's solos and tried to assign fitness levels to the solos' constitutive units while my index finger was ready to hit the G key and my thumb was ready to hit the B key, I found it extremely difficult to discern musical structures that I liked. Consequently, I ended up hitting the B key time after time. I did so less because I hated what I heard and more because I could hear very little that made distinct enough musical sense. After I finished assigning fitness levels, David opened a file called "PopStats," which displayed different statistical data on the currently trained measure and phrase populations. He then compared the data to the statistical data that the computer compiled about the population that he had trained before me. David said:

> Our average fitness for measures is almost minus 2 now [David highlights in black the fitness level -1.91; see fig. 1.2], so it went down. So you were suitably critical. It looks like we got one that is a minus 20 [David refers to measure 9]. That's not going to hang out for very long. That's the worst one by far. Here's one that's a minus 9 [David refers to measure 12. He scrolls down the file until he reaches the statistical data on the phrase population; see fig. 1.3]. OK, average phrase fitness is minus 3. Again, kind of going down. If we look through the distribution, there are some that are—here's one that is a minus 13 [David refers to phrase 1 or 17]. Yeah, it's a pretty negative population. And for a soloist that to my ears didn't sound that bad.

Second, although using a human being's subjective judgment to assess fitness is not considered as gross a violation of the genetic-algorithms framework as designing "intelligent" mutation operators (Helmreich 1998a:44),

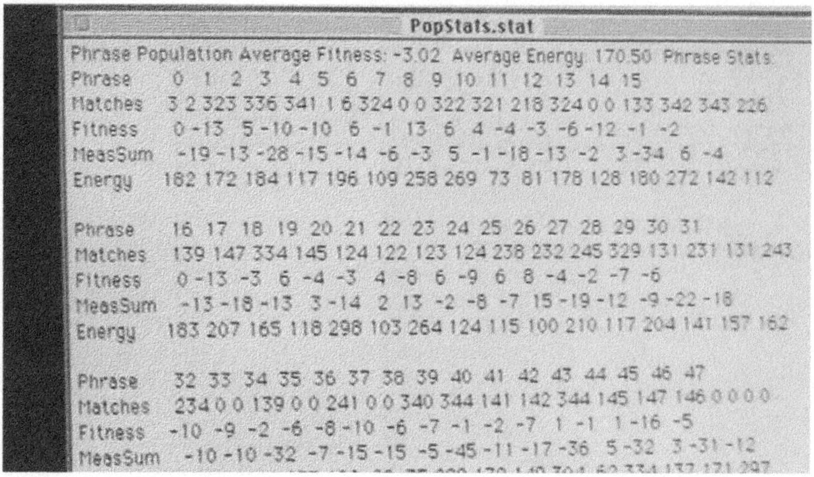

FIGURE 1.3. A file displaying different statistical data on a trained phrase population.

using subjective judgment in a system that is based on this computational framework and that is specifically meant to function in the domain of creativity and aesthetics produces specific challenges. First and foremost, human aesthetic judgment as a source of fitness evaluation has the inconvenient tendency to shift together with the objects that it is supposed to evaluate, especially when those objects are a work in progress, such as numerous generations of a phrase population that are played in real time over long solos. During our joint training session, after we had trained a soloist, David said:

> So I started by being positive but then I said, "Wait, they are all sounding this way." So I started backing off and then I thought—"Well, I really don't prefer this specific phrase"—so I had to make myself more discerning, quicker than usual. . . . What will typically happen sometimes is there would be a nice figure that you like early on in the training and you reward it and then it just proliferates and it becomes annoying because there are so many of them and then you start breeding them out. You often change your mind as you go. It's a phenomenally noisy process.

David explains that during a training session, his aesthetic judgment of the same phrases or measures might change as a function of how many times these phrases or measures appear as a result of the fitness levels he had assigned to them earlier in the training process. A phrase that he liked earlier in the training process might annoy him later in the process because of its ubiquity, which is itself the result of David's earlier preference for it. The result is "a phenomenally noisy process" that prompts the question of what the advantage is of using evolutionary computation rather than rolling dice to generate musical phrases. Aesthetic judgment and standards can thus vary not only across cultures, historical periods, and different members of the same culture (Boden 2003:10; Jordanous 2019:221), but in the course of the evaluation of the same objects made by a single person, too.

David's De-Deification

Owing to the different challenges and frustrations that David experienced in the process of designing his system, he eventually gave up hope that he would be able to personally breed away every single lick that annoyed him in the real time of human- or machine-based improvisation by a mere stroke of the B key. And because he was also unable to program a fitness function that could evaluate the aesthetic fitness of individual phrases, he decided to reconfigure his system in a way that would entirely obviate the need to have any kind of real-time evaluation.

First, instead of programming the system to produce the first generation of phrases by generating random bit strings of the appropriate length for each phrase, David created a database of phrases that he took from a published method book containing hundreds of conventionalized jazz licks arranged according to different jazz styles and presented in the context of different common chord progressions. He programmed his system to generate the first generation of phrases by randomly selecting sixteen licks from this database and then creating thirty-two additional child phrases by means of a crossover of those sixteen licks, totaling forty-eight phrases.

Second, David used a pitch-to-MIDI interface that made it possible for him to seed his system's phrase population with his own musical ideas in the real time of their joint improvisation. He would determine in advance the order in which he and the system would improvise, one after the other, during the joint improvisation. The phrases he then played on his trumpet into a microphone once his turn to improvise came were mapped in real time onto the system's chromosome representation. David explained the way his phrases seeded the system's phrase population:

> Originally, I was thinking: "OK, when I play a solo, every measure and phrase I play that is good—add it to the population, breed it in there." But eventually I went in another direction. What I ended up doing is, it [the system] would take every measure that I played and find a measure in its personalized notes that is as close as possible to the ones I just played and then do an intelligent crossover there and then take one of those "kids" and put it in there instead. And that basically drives that soloist in the direction of the tune I play and

my solo, if I take a chorus before it does, and that's very cool. Because even though it starts with those original licks in its database [the licks taken from the method book], it's still going to mess it up and bias it toward what I play.

Third, whereas in the system's first version the recombination of the chromosomes of phrase individuals took place at a random crossover point along each phrase's bit string, in the system's later version the recombination took place by means of an "intelligent" crossover, as David mentioned in the last quote. Crossover points could now occur only at the boundary between two measures within a four-measure individual phrase. Furthermore, the crossover points that are selected are those that will produce children that most closely preserve the horizontal interval between notes at the crossover point in the two parents, that is, the interval between the last note before the crossover point and the first note after the crossover point. These changes make sure that the recombination of parent licks, which are conventionalized stylistic building blocks, will not compromise those licks' musical meaning but rather produce child phrases that will carry this meaning to the next generation.

Finally, David made changes in the kind of intelligent mutation operators that his system uses to produce musically meaningful diversity. These mutations operate on measures that have already been played at one point in a given solo. They include the random transposition of a measure up or down one to three steps; the inversion of a measure, that is, playing it upside down; a retrograde, which reverses the events in a measure; and a mutation, which both inverts and reverses a measure.

To tease out the broader implications of the changes that David made in his system, consider the following exchange that took place between the two of us in the lab after the system's later version improvised on the standard tune *Four*:

EITAN: So this version of the system doesn't have a fitness function, right?
DAVID: Right. I still use the training version [the early version] from time to time. I have to use it for things in odd meters like 7/4 or 5/4 because I don't have databases of licks in these time signatures. The book of licks that I used doesn't have licks in these odd meters. And, of course, I can still train soloists for the common time signatures, but I never bother anymore.
EITAN: OK, so would you say that in the new version, without the fitness function, there is still some kind of evolution? Because earlier you said that it evolves.
DAVID: That's a good question. When I presented the system in different conferences, this is when people started saying, "You're cheating. There is no fitness." But I would argue—and this is what I do argue when I get

cornered on this—that it [the system] follows the evolutionary paradigm: it has a positional representation, so there is genotype-phenotype mapping—this is very clear; it does crossover; it does mutation. But the one thing it doesn't do is generational evolution. But that doesn't matter given the constraints that I've got on it, where generation zero are proven licks [from the method book] and it's doing an intelligent crossover—either blending two licks or phrases together, one or more measures from one, one or more measures from another, and I'm going to get then another individual from the phrase population that uses the same measures in a different way.

EITAN: But would you say that the new phrases are more evolved?

DAVID: I would argue that that is evolution! It doesn't happen to require fitness because if all the measures are good from the get-go, and if we do intelligent crossover and avoid huge jumps at the crossover point, it's gonna tend to sound good. It's going to blend those together. Then, when it was soloing there, every time a measure was played, regardless of the phrase played, after it plays that measure it goes and mutates it, so the next time it plays it, it will be a different measure. And because all the mutations are also intelligent or musical, the result will sound good.

EITAN: But from what I understand, in evolution there is something that decides if one thing is better than another thing.

DAVID: Yeah, it decides, and my whole point is that if your initial population is all good, and your crossover and mutations do good things, then I don't have any need for fitness.

EITAN: Because everything is good.

DAVID: Right.

EITAN: But the question is whether it evolves.

DAVID: I would argue that it does evolve. It is not a guided evolution and that's what motivates people in those conferences to throw it [his system] to the garbage bin.

EITAN: Is what you are describing something that can also be a devolution?

DAVID: What's worse or better anyway? Because I don't know how to write an algorithm that can decide that.

EITAN: Well, I can tell if something is better or worse.

DAVID: Yeah, OK, but I never hear things that are worse. That's the thing.

EITAN: Can you sense any improvement in the solos that the system generates?

DAVID: Well, those little evolutionary adventures only happen for the duration of one tune at a time.

EITAN: OK, so do you hear any improvement or evolution in the solos that the system generates for the duration of one tune?

DAVID: No, not over time, because if you give the intelligent mutation and crossover operators good stuff—the initial musical population of licks [derived from the method book]—you get back good stuff.

This exchange, which has the flavor of a cross-examination in court, clarifies a number of important points about the changes that David made in the system. First and foremost, after not being able to realize his dream of breeding away annoying licks by personally embodying the fitness function and assessing successive generations of phrases in the real time of improvisation or by programming this function, David decided to make sure that no annoying licks would be produced in any of the future generations of phrases that his system produces to begin with. He made sure that each generation zero would consist of conventional, that is, already musical, licks; that all generations would be seeded with his own, already musical licks; that the recombination of the chromosomes of these phrases would maintain their musicality; and that the mutation operators would be musical, too. In effect, David decided to forgo the fitness function altogether—to forgo having a feature that would determine the fitness of each phrase as it is produced—and instead to determine in advance the fitness of all future generations, that is, what the population of "good" solutions or musical phrases and licks would essentially look like.[9] If genetic algorithmists tend to exhibit a predilection for a mix of control and surprise by setting out the initial parameters of an artificial world and then enjoying the drama and suspense of the emergent and unexpected phenomena to which these parameters give rise (Helmreich 1998b:75; Turkle 1991:235), David exhibited a predilection for complete control in both versions of his system.[10] He first tried to exercise complete control by personally performing the fitness function during his system's production of successive generations of phrase populations so that he could eliminate any unpleasant surprises in the form of annoying licks whenever they occurred. He then shifted to exercising complete control by designing the system's basic parameters in ways that would ensure that such unpleasant surprises would not be produced, to begin with. To this he added a kind of safety measure in the form of seeding the phrase populations with his own ideas during the system's improvisation.

Second, and equally important, the price that David had to pay for making sure that his system would not produce unpleasant surprises when it generated future generations of phrases is that he also eliminated the possibility that his system would produce pleasant surprises: musical phrases that are significantly different in style from, and more aesthetically pleasing than, the generation zero of musical phrases and the general stylistic parameters that he encoded in the mutations. Only at the end of my cross-examination did

David acknowledge that while he "never hear[s] things that are worse," he also never "hear[s] any improvement or evolution in the solos the system produces for the duration of one tune . . . because if you give the intelligent mutation and crossover operators good stuff—the initial musical population of licks [derived from the method books]—you get back good stuff." When later in our session I asked David whether his system ever surprises him with its improvisations, he responded:

> No. Well, occasionally it will do something where I will be: "Hey, that was really nice," you know? I guess it cannot get any better than that. . . . A lot of its solos are fairly even in their contours, and if you take bits and pieces of them they are not recognizable. That's really what I want. What I don't want is that you'll hear a recognizable lick followed by a recognizable lick followed by a recognizable lick. . . . It may never be brilliant but it will never sound bad. And I was looking for a competent sideman, you know. I never want it to sound wrong. If it occasionally sounds brilliant, then terrific—to someone's ears, at least. But I never want it to sound incompetent. As long as it's competent I can live with it as a sideman and then I might only be pleasantly surprised.

David is content with knowing that his system will never or only rarely pleasantly surprise him with its improvisations, as long as he can be sure that it will never unpleasantly surprise him by playing repetitive licks or theoretically wrong notes.

David's willingness to accept his system's stylistically predictable improvisations modeled on his own style, and his description of his system as a competent sideman, suggest that the musical interaction between David and his system is not one between equal partners. On the one hand, this might cast a shadow on his achievement, inasmuch as contemporary definitions of "artificial improvisers" or "live algorithms" stipulate that such improvisers "must be able to collaborate with others. Both human and machine contributors must have equal status. This means that contributions seem equally valuable as independent statements and as offerings to the collective. Demonstrable equivalence is a prerequisite of true *interaction*, in the non-technical, commonly understood social sense" (Young and Blackwell 2016:507–508). On the other hand, David's interaction with and description of his system suggest that such definitions of artificial algorithms are problematic in that they are decontextualized and prescriptive. Artificial improvisers do not spring from a social void; they are frequently designed in response to specific social problems. David designed his system in view of a very specific predicament that he had experienced in jam sessions: the atomized and repetitive playing of human musicians, which made them incapable of participating in a musical interaction with fellow musicians. Consequently, he designed his system to exhibit specific dimensions of interactional competence, as I will describe in greater detail in the next chapter. It was more important for him to make sure that his system knows how to take into account his real-time improvisations and meaningfully respond to them—also in real time—than that it knows how to produce genuinely creative ideas that could surprise him. Furthermore, the design of artificial improvisers is frequently informed by existing institutional models or forms of musical interaction. David's characterization of his system as a competent sideman invokes the idea of a jazz band with a bandleader. Contrary to the prevalent myth about jazz as a radically democratic form of music, many of the canonic strands of jazz were the product of hierarchical organizational forms—from big bands to small groups—led by powerful bandleaders who set the tone (no pun intended) with respect to many of the group's aesthetic and organizational dimensions (Berliner 1994:418–419; Givan 2022; Wilf 2014a:86). Against this backdrop, David's description of his system as a competent sideman perfectly aligns with an existing institutional form of organized musical interaction in the cultural order of jazz. If demonstrable equivalence has not always been "a prerequisite of true *interaction*, in the non-technical, commonly understood social sense" (Young and Blackwell 2016:507–508) in the jazz world, the demand that artificial improvisers exhibit it should be qualified.

A third point that is noteworthy about my exchange with David is that his programming choices have made him persona non grata in evolutionary-computation academic circles. In the eyes of genetic algorithmists, his decision to forgo the fitness function has left him with the empty shell of the evolutionary-computation framework.[11] David responded to this rebuke by reorienting his goals. He explained:

> It's [his research focus] started being less about technology and more about the music, and ultimately this is the kind of reorientation technologists have to have if their technological work is going to succeed in an artistic way. It has to be about the art, not about the technology. That's what most of the computer science programs that I see, the genetic algorithms that play stuff, those people in France that are breaking their butts to come up with a fitness function, an algorithmic fitness function so that it won't be subjective—Hello! It's about being subjective! And then for me it was kind of a gradual slide into— it's a music system, it's not a genetic algorithm. . . . And if I have to break evolutionary computation to play the kind of music I like to play, for me that's an easy choice to make. But for a computer scientist, that's cheating. I was going to the GECCO conference [Genetic and Evolutionary Computation Conference], I was doing a tutorial there on evolutionary music for a few years, and I

would get people come up to me and say: "You shouldn't be here because you do not have a fitness function, you're cheating." You know? It was a bunch of engineers interested in optimizing another 10 percent of their algorithms. . . . They could live with the fact that my mutation operators are intelligent and not stupid, even though in a normal genetic algorithm the only thing that is supposed to be smart is fitness. Everything else is supposed to be dumb. But they couldn't live with the fact that there is no evaluation. That's why I quit going to those conferences.

When David's attempts to design a digital jazz improviser in accordance with the evolutionary-computation framework ran into difficulties, he decided to prioritize his aesthetic preferences—"the music" and "art"—over strict adherence to the precepts of this computational framework. He did so to avoid the fate of other programmers who are engaged in similar projects and who, by insisting on sticking to the evolutionary-computation framework, have produced what David considers to be unsatisfying results.

David's argument that his project is more "about the art, not about the technology" should be qualified, because he time and again performed or mediated this argument by means of his system's technological infrastructure, thereby making his project about the technology as well, but in a very specific sense. For example, during our interactions he frequently emphasized and made fun of the fact that the computer he uses for his system—an Apple Macintosh PowerBook 180, which was sold to the public from 1992 until 1994—is so out of date. Thus, when we first met for dinner, I turned off my phone—an old Samsung flip phone. While doing so, and because I was self-conscious about it in view of David's identity as a computer scientist, I mumbled that it was ridiculous that I was still using such an old phone. David responded, "It works! Wait till you see what my system is running on!" Later during dinner he described, and seemed to relish, people's reactions to his system's antiquated technological infrastructure:

> I went to London and set up to play a concert and I pull my tone generator out and this guy looks at it and says: "Oh, my god, I used to have one of those!" And then I pull the computer out. "Oh my god! I used to have one of those too. Does this still work?" [David smiles as he imitates his amazed interlocutor.] Or I play in the Computer History Museum for the board of trustees of our university, in Palo Alto, and I'm thinking: "Wow, they're going to think I'm ripping off the exhibit." And there was one [another Apple Macintosh PowerBook 180] in a glass case. It's embarrassing, but it is kind of fun. And I have fun saying it's a 14 meg system with a 120 meg disk and a 32 megahertz processor and it does this. And it runs on 2 meg. And the executable file is half a meg. [David laughs.] There is a subversive quality to that.

On the one hand, David minimizes the role of the technology in his overall project, thereby making his project "about the art." On the other, technology plays an outsized role in his performative minimization. David time and again emphasizes and displays the very modest capacities of his technology and notes that there was even a specimen of it enclosed in a glass case at the Computer History Museum in Mountain View, where the concert he was giving took place.[12]

The differences that David had with the community of genetic algorithmists eventually led him to stop attending their conferences. His initial decision to withdraw from the company of his fellow musicians in jam sessions and to develop his system was thus followed by a decision to withdraw from the company of his fellow researchers in specific academic conferences. Significantly, as I describe in the next chapter, his gradual withdrawal from these two contexts of sociality with humans coincided with the tightening of his sociality with his computerized system. This shift, in turn, rather than causing his music interactional abilities to atrophy, actually allowed him to cultivate and perfect the skills that in the jazz world are considered to be essential for musical sociality. This outcome represented a significant achievement, given the goals that had originally motivated David to design his system. At the same time, it had specific downsides, too.

2

An Island of Interactivity in an Ocean of Nonreactivity:
The Trade-Offs of a Made-to-Order Artificial Musical World

My Kind of Artificial World

During my long playing and training sessions with David and his system, I noticed that David would occasionally smile while he listened to the improvisations that his system produced in real time in response to his or to my improvisations. At first I thought his reactions reflected the kind of paternal amusement that many inventors display regarding the behavior of the artificial systems they "father" (Helmreich 1998b:120). Indeed, whenever I teach my seminar on human-machine interaction, I discuss this feature with my students in the context of a short film in which the cybernetician William Grey Walter demonstrated the results of his pioneering work on electronic quasi-autonomous robots in the late 1940s.[1] The two robots that he designed, which he called Elmer and Elsie and were also known as "Grey Walter's tortoises" because of their shape, slow movement, and small dimensions, were equipped with light-sensitive sensors that guided them to sources of light and touch-sensitive sensors that led them to make avoidance maneuvers whenever they hit an obstacle. In the film, when the two robots wander around in the room and display complex behavior as a result of the interaction between these simple parameters and the robots' environment (that includes each other, too), the camera occasionally focuses on the faces of Walter and his wife, who look down at the robots and smile as if they were parents enjoying the mischievous behavior of their offspring.[2] The circusy music that accompanies the video, and the fact that Walter places obstacles in front of the robots with his leg while they are moving, contribute to the impression that the two humans' smiles are not only paternal but also patronizing, in that they are an expression of a hierarchy in the context of which the more complex designer tests the simpler (yet complex for machines) behavior of his creation and finds amusement in it.[3]

However, when I asked David to describe what he actually liked about his system, I learned that his pleasure was multidetermined and included a number of dimensions that had very little to do with a sense of his own musical superiority and more with an appreciation for something he considered to be his musical equal and, in some ways, superior. To be sure, what started as David's attempt to design a jazz-improvising digital system by means of a computational framework predicated on the promise of emergence and surprise has gradually given way to the design of a system that David prized for the different kinds of control and predictability that it provided him. At the same time, when David described the ways in which his system represents a more controlled and satisfying musical environment than the world of human musicians in which he had participated and the differences between the two musical contexts what came to the fore was that he described the system's musical capabilities as equal and in some instances as superior to his own.

First, to reiterate, David liked the fact that his system generates improvisations that mirror the stylistic dimensions of his own improvisations. He compared the benefits of this mirroring with his disappointing experiences in jam sessions. "The system will always be playing inside the harmonic changes, and at heart I am an inside player," he told me. "Many times in a jam session, if you are out there and the tenor player is playing with you on this tune, and he's honking and squeaking and playing far-out stuff and you're trying to sound like the early Miles [Davis]—you know, it doesn't work very well. There's no conversation. It's two guys talking about different things and past each other." David describes a jam session in which his desire to sound like Miles Davis at the early stages of his career, when Davis played lyrically and inside the harmonic changes, clashed with a saxophonist's desire to play "far-out stuff" that lay outside the harmonic changes and to use sound effects such as squeaks. Against this disorienting backdrop, David's system offers him a sort of haven, as he explained to me:

> The system plays inside the harmonic changes—it's basically a diatonic player. So it's going to play inside, it's going to sound like I'd like to sound, because I also play inside. It plays eighth-note multiples, which I tend to [play] too. Also the intervals—they are relatively small: it doesn't hop around a lot. That's also my aesthetic. It doesn't play loud or fast and, you know, in your face, and I don't either. I don't like to play like that. So we're a good match for each other. It's like playing with myself. [David pauses for a few seconds before adding:] But in terms of actual licks, it doesn't have any of my licks other than in the moment when we are trading fours. That's one of the reasons I decided not to take four-bar phrases that I played and stick them into its population because I really didn't want it to mimic what I was doing at that level. I wanted it to be

influenced by what I do, to take vestiges of what I did, . . . where you can hear things like what I did, but I didn't want it to just copy what I did because that wouldn't be truly interesting.

David describes all the different ways in which the playing of his system reflects and is aligned with the aesthetics that informs his own playing. Although David is careful to emphasize that the system does not play his exact phrases and licks back to him—because it mutates them—he expresses his pleasure at hearing in the system's improvisations "vestiges of what I did, . . . where you can hear things like what I did." David reveals his autopoietic predilections when he says, "I wanted it to be influenced by what I do," and when he argues that playing with the system is like "playing with myself."

Second, David argued that his system knows more tunes than most human musicians do. "Most of the people I played with don't know as many tunes as I know and as it [his system] knows. Actually, I don't think that person even exists. Because right now I've got about three hundred tunes that I play with the system on a fairly regular basis. Three hundred plus, actually. I have around three hundred tunes that I would consent to play for someone and then one hundred or so that are either retired or they are in my rejects folder—tunes that never really came together." David prizes his system's vast knowledge of tunes because one of the first decisions musicians must make in a jam session is deciding which tune they will improvise on together. Consequently, in order to function well in a jam session, musicians must master a large enough repertoire of standard tunes that they can play on a moment's notice. This is why the title of a recent ethnographic study of the jam session as a social institution is *"Do You Know . . . ?" The Jazz Repertoire in Action* (Faulkner and Becker 2009), for the most important question one is likely to ask or to be asked in a jam session is, "Do you know [this or that tune]?" David can be sure that his system will know every tune that he knows and wants to play because he taught them to his system.

David tacitly drove this point home for me when at one point during our playing session with his system he asked me, "What kind of tune do you want to do now? How about a waltz? Do you know the changes to *Booker's Waltz*?" "No," I said. "How about *Charade*?" "I don't know that one," I confessed. "*Emily*?" David tried another tune. "Sorry to disappoint you, I don't know those," I mumbled with increasing embarrassment. "It's all right," David said. I pointed at the computer screen, which displayed the files of the different tunes he had taught his system, and said: "You have the entire jazz repertoire here." "Yeah, I got at home a whole bookshelf with books of tunes and stuff. What about *I Dream of Jeannie*?" David tried one last time. "I don't know that one either,"

I said. "OK, let's do *Days of Wine and Roses*," he said. "Sure," I replied with relief. The reason David did not ask me whether I knew *Days of Wine and Roses* is that he assumed I did—it is a well-known standard played frequently in jam sessions. He only suggested it after his attempts to have me play lesser-known tunes had failed. For someone who enjoys playing tunes that are not regularly played in jam sessions, David's system was clearly a better musical partner than I was. But that was not the only reason it was better: not only can David be sure that his system will know every tune that he knows and wants to play, but he can also be certain that he will not be asked by his system to play any tune he doesn't know, and thus he will not find himself in the same uncomfortable position in which he inadvertently put me during our playing session. This last guarantee is not to be underestimated, for a musician who does not know a tune that is called out during a jam session can immediately lose face.

Third, there are many ways in which musicians can know tunes. Some musicians might know a tune's harmonic progression but not its melody. Other musicians might know a tune's melody but not its harmonic progression. Consequently, even when a group of musicians decides to play a specific tune in a jam session, the result can be less than satisfying. In contrast, as David explained to me, he can be sure that his system will be able to play well enough any tune in its database:

> Some of it is executing the tune correctly, you know, the fact that it hits all the notes and all that. I mean, everybody makes mistakes and you put up with it. But *it* doesn't really make mistakes. Some of the repertoire, the tunes that I do, you know, it would take a long time for a bunch of players to work this out. Some of the tunes are just hard. And then you have the bass player who wants to mess it up sometimes, and the piano player who comes [in] at the wrong time. But with the system it's easy to set up, because once there is a score it's going to work. I have total control, and it's going to come out the way I want it to come out, and if it doesn't, I just don't play it.

David's frustration with his fellow musicians stems from the fact that not only might they not know a given tune well enough to execute it, but they might also decide to be creative with it—"to mess it up," as David puts it—by, for example, playing alternative harmonic changes. In contrast, David's system provides him complete control over the execution of tunes. David added that because having a system that executes the tunes well is important to him, he has never been interested in experimenting with a system that plays an acoustic instrument, because such a system would be more likely to make errors of execution and to malfunction due to its mechanical complexity.[4] In addition, such a system would be limited to playing one instrument, which would be a problem as far as David is concerned. "It's a challenge from a mechanical engineering standpoint," he told me, "and I don't really care. I'm not really interested in that. I am more interested in the notes that my system plays and that they are executed correctly, not that they are executed on an acoustic instrument. Plus, my system can play any instrument that I choose. There is a physical modeling card in there—a tone generator. It's ancient but it works. It does an incredible job."

Fourth, David argued that because of its design, his system will always generate improvisations that accord with the harmonic progression of whatever tune it improvises on, as opposed to human players who often err in this respect. There are two aspects to the system's ability to accomplish this feat. First, the system has an arsenal of eighteen different chords and their respective scales, which allows it to handle the different harmonic progressions of different tunes. David explained:

> Its harmonic knowledge is very robust, so it can work with very different kinds of tunes. And so the scales are always going to be reasonable choices. I have different kinds of scales mapped to eighteen different chords. Because when I do a tune, when I want to do a tune, well, if I don't have a specific chord but another one [in the system's arsenal of chords] is pretty close and I plug it in and when it gets to that space it starts sounding weird, I can just— "Oh, OK, let's try another one. Ah, this one is not working, I better add a new chord. Gotta do this one now." So that's how it got to something like eighteen different chords.

Second, because whatever musical phrase the system plays is always going to be mapped onto the specific scale that matches the chord the system improvises on, the system's playing will always be theoretically sound. "It's going to get it right the first time every time because it cannot not get it right," David said. "It cannot play a theoretically wrong note. Any melodic contour [i.e., phrase] that it works with is always going to be mapped correctly to the harmonic changes and it will play notes that are reasonable [harmonically], and people don't do that that well, which can be good and bad." David acknowledges that playing harmonically incorrect notes can be a good thing in some situations when it is done intentionally, in that it can create productive tension (what jazz musicians commonly refer to as playing "outside the changes"). However, because, as he acknowledged earlier, he prefers to not be unpleasantly surprised even if it means that he will never be pleasantly surprised, and because he is not a player who enjoys playing outside the harmonic changes anyway, David prefers having a system that "cannot play a theoretically wrong note."[5] He concluded: "If the criterion is to take a tune and play an original solo on it and hit the changes and sound good, then it can play a lot better than a lot of people."

Last, having the system at his command spares him the need to deal with the logistics of organizing a playing session with human beings: "The logistics are that the people you'd like to play with are going to have a life, so scheduling them for a gig will be difficult. I had a trio where we missed some gigs. This trio I got, there were only the three of us but we missed a gig once because the guitar player couldn't get ahold of the bass player in time to confirm the date, and they gave it to somebody else. You know, it's no big deal, but for me, with this system, if I'm available, it's available, and that makes life easy."

David's system thus offers him a musical companion that will always play in his style and according to his aesthetic preferences, can play any tune that David knows and wants to play, will know these tunes well and execute them correctly, will consistently improvise on those tunes in a theoretically correct way, and will always be available whenever David feels like playing. In contrast, the musicians David played with, especially in jam sessions, often played in styles that were diametrically different from the one in which he prefers to play, did not know many of the tunes he wanted to play, could not execute them properly or took liberties with them even when they did know them well, played theoretically incorrect notes, or were simply unavailable.

David's description of his appreciation for his system's capabilities shows that although one of the reasons that motivated him to design his system was his desire to have a musical partner that would play in a less predictable manner against the backdrop of the fact that he felt his fellow musicians played in highly repetitive and derivative ways, his goal was not to increase all forms of contingency that might be present in a musical interaction. David used the stochastic or pure contingency afforded by the computer to increase up to a point its social contingency as a musical partner, one whose intentions and actions can never be fully anticipated in advance with complete accuracy. At the same time, he made sure that the computer's playing will always make musical sense, thus reducing the semiotic contingency associated with the potential unpredictability of meaning and interpretation of any action, and that the computer always perfectly executes whatever it is playing, thus reducing the performative contingency that has to do with the execution of an action by a participant.[6]

A scholarly framework that comes to mind given the artificial microworld that David designed is the one developed by scholars who have studied the emotional dimensions of human-machine interactions, especially interactions with relational or interactional artifacts. These scholars have argued that in the contemporary moment, a growing number of people—adults and children alike—are inclined to turn to digital relational artifacts in search of a safe and risk-free environment that can be trusted and that can provide them with a corrective to their disappointment with their fellow humans, whom they perceive as unpredictable, as having desires and agendas that oppose their own, or simply as incompetent (Turkle 2011:66). David's artificial microworld also brings to mind the argument made by such scholars about the new risks that replacing humans with relational artifacts might entail:

> [During normal socialization] we learn to tolerate disappointment and ambiguity. And we learn that to sustain realistic relationships, one must accept others in their complexity. When we imagine a robot as a true companion, there is no need to do any of this work. The first thing missing if you take a robot as a companion is *alterity*, the ability to see the world through the eyes of another. . . . If they can give the appearance of aliveness and yet not disappoint, relational artifacts such as sociable robots open new possibilities for narcissistic experience. . . . Those who succumb will be stranded in relationships that are only about one person. (Turkle 2011:55–56)

In her study of repeat gamblers in Las Vegas and the design of digital gambling machines, Natasha Dow Schüll (2012) makes similar arguments by drawing not only on Sherry Turkle but also on scholars of video games and child psychology. First, she argues that for repeat gamblers, "the gambling machine is not a conduit of risk that allows for socially meaningful deep play or heroic release from a 'safe and momentless' life (to use Goffman's phrase), but rather, a reliable mechanism for securing a zone of insulation from a 'human world' [they experience] as capricious, discontinuous, and insecure" (13). Second, this "personal buffer zone against the uncertainties and worries of [repeat gamblers'] world" (13) is based on a kind of fusion between the gambler and the machine, which is a result of the fact that gambling machines are designed to closely match and adapt in real time to individual gamblers' gambling preferences, dispositions, and styles (151–152): "immediacy, exactness, consistency of response: the near perfect matching of player stimulus and game response in machine gambling might be understood as an instance of 'perfect contingency,' a concept developed in the literature on child development to describe a situation of complete alignment between a given action and the external response to that action, in which distinctions between the two collapse" (172). "Over the course of repeated play," a gambler might come to feel "that a technological object is an extension of his own cognitive and even motor capacities"—a feeling that "replaces a sense of the machine's alterity"—because gambling machines are designed to tightly match the gambler's actions, rhythm, preferences, bodily disposition, and more (174). Third, whereas Turkle points to the "new possibilities for narcissistic experience" that a user's relationship with sociable robots affords (Turkle 2011:56), Schüll argues that "the operational logic, capacitive affordances, and interactive rhythm of the modern gambling machine . . . make[s] it a particularly expedient vehicle for retreat into the 'functional autism' of perfect contingency" (2012:173), which finds expression in one's preference for "sameness, repetition, rhythm, and routine," and for the predictability of "another's perspective or intentions" (172). Such a stance is overall an "antisocial" one, because the tight coupling of human and machine often leads gamblers to prefer it over interaction with people who are endowed with their own agendas, preferences, and predilections (173).

On the one hand, David's case aligns perfectly with these observations and predictions insofar as the artificial world he built is predicated on the aesthetic dimensions of his own playing, and given the fact that he built it to avoid the different kinds of disappointment that he had experienced when he played with human musicians, and to have complete control over every aspect of the playing situation. On the other hand, it also complicates these arguments and predictions. Turkle has argued that "dependence on a robot presents itself as risk free. But when one becomes accustomed to 'companionship' without demands, life with people may seem overwhelming. Dependence on a person is risky—it makes us subject to rejection—but it also opens us to deeply knowing another. Robotic companionship may seem a sweet deal, but it consigns us to a closed world—the loveable as safe and made to measure" (Turkle 2011:66). As I now will show, the highly controlled musical environment that David designed by means of his system became a condition of possibility for his ability to cultivate the musical skills that in the jazz world *are* considered necessary for "deeply knowing another." Furthermore, David was able to take those skills back with him to, and apply them in, musical interactions with players whenever he felt like participating in them. Crucially, those musicians noted the improvement in David's ability to be open to their playing in real time—the result of his prolonged musical interaction and sociality with his digital system.

A Very Tight Conversation

The core of the pleasure that David derived from playing with his system is anchored in a specific form of musical interaction that is prevalent in jam sessions. This form of interaction, which is known as "trading fours," typically takes place immediately after each musician improvises a number of choruses over the harmonic progression of a tune. The different musicians then improvise on the tune a few measures each, one after the other, and in response to one another's improvisations. The art behind this practice is to take what another musician just played and use it as a resource in one's own improvisation. These back-and-forth contributions that refer to, continue, and develop each other are the reason that trading fours is commonly thought of in the jazz world as a kind of conversation (Berliner 1994:668–672; Monson 1996:77–80). Recall that one of the main reasons David decided to design his system was that in most of the jam sessions he had attended, musicians were more focused on their own improvisations and were oblivious to what their fellow musicians played, with the result that, in his words, "there's no conversation. It's two guys talking about different things and past each other."

In response, David programmed his system to have a trading-fours interactive mode based on its modified evolutionary computational architecture. This mode became the cornerstone of his musical interaction and sociality with his system. When David plays a four-measure improvised contribution in this interactive mode, the system immediately maps it as best as it can onto a chromosome-structure representation. It stops "listening" in the last thirty-second note of David's improvisation (a four-measure phrase in 4/4 consists of thirty-two eighth notes), mutates the chromosome structure of David's improvisation, and then plays the result as its own four-measure improvised contribution. As I discussed in the previous chapter, the mutation operators are well-known compositional techniques such as rotation, retrograde inversion, retrograde reversal, and transposition.

The algorithmic architecture that David designed for this interactive mode represents an attempt to follow the jazz community's normative ideal of what a musical conversation between players should look like. David's system can formulate musical responses to David's playing, which more closely align with this ideal in specific dimensions, compared with the responses that human musicians might be able to formulate to his playing. When I asked David whether, as a result of trading fours with his system for so long, he has learned to formulate musical responses to his system's improvisations in real time by using the same kinds of compositional techniques that underlie the system's mutation operators, he responded immediately:

> No, that's absolutely not the case. The things that it [the system] does to develop what it hears me play have nothing to do with the things I do to develop what it plays that I hear. I mean, if it chose to—if the dice came out that way—it might take my four-bar phrase and play a retrograded inversion of the entire four bars and hit all the changes for the next four bars, and there's

FIGURE 2.1. The author improvising together with David's system, using one of David's trumpets, while David is seated, listening.

no human alive who can do that. There's no way! And it's going to sound good because the notes it hits are going to be reasonable. And if the melodic contour is decent, playing it upside down and backwards is going to sound nice too, because at a micro-level the intervals are going to be nice, so it will sound good. There's no human who can do that in the heat of the moment. As I kept adding to it and listening to it better and adding more mutation operators and more things that it can do and tweaking it, I got more respect for it as a player. And then when I compared what it does on a tune to what people do in the jam sessions that I go to, if that's the bar—the jam session—it blows them out of the water, sorry.[7]

During my fieldwork with David I had the chance to trade fours with his system on a number of tunes using one of David's trumpets (see fig. 2.1). I observed the system's superb ability to take vestiges of what I had just played and use them in a transformed and musically meaningful way in the improvisation that it produced immediately after I finished my four-bar improvisation. What was noteworthy was not just that the system managed to produce this feat, but that it managed to produce it virtually every time it responded to my playing. Whereas many human players can take their interlocutors' improvisations and meaningfully incorporate them into their own responses, very few of them can do so every time and in a consistent manner.

Although David's system represents an attempt to follow a jam-session conversational ideal that he could never hope to reach and that, based on his experience in jam sessions, most of the players he played with did not even try to reach, he argued that being around and musically interacting with a system that does reach this ideal has significantly improved his ability to function well in a musical-conversational context:

> You know, my boss told me a couple of times: "You have to be a different player after you've been doing this for all these years." And I've noticed it, too, because when I came out of the cold and started playing with people in the last few years, people kind of dropped jaws a little bit and said: "Wow, when did you become this good?" There's this guy, a retired faculty member. He used to play bass with me a lot. We used to play in jam sessions for a few years and he hasn't heard me play for a while, and when I went to one of these jam sessions in town he was there and he said: "Wow, you sound good, what have you been doing?"

When I asked David to reflect on the nature of the changes in his playing that his colleagues noticed, he said, "Well, I'm smoother, I hear the tunes better, I hit more changes, I am more lyrical, I play better melodies, I'm much more spontaneous, but also much more inside [the harmonic changes]—I'm more inventive within the constraints of playing in the changes. I don't get lost that much. So I really can hear what it's doing and—'Oh, that gives me an idea,' and off I go. One important thing I found is that when I play with people now I'm really good at catching what someone else just played and putting it back at him." David's description of how his playing has changed following the long period of interacting with his system is noteworthy. First, David argues that the musicians with whom he played after the long period in which he limited himself to playing mostly with his system noticed radical and positive changes in his playing. His use of the phrase "coming out of the cold" is reminiscent of a specific trope in the cultural order of jazz, which has to do with famous musicians who decided to eschew playing with other musicians for a while in order to hone their skills in private and then returned to the jazz scene completely transformed and able to display in their playing higher levels of jazz artistry.[8] Second, and most important, David argues that because he had so much experience trading fours with his system, he has become much better "at catching what someone else just played and putting it back at him"—that is, he has become better able to engage in a meaningful musical conversation with his fellow musicians in real time according to the jazz community's interactional normative ideal. His long experience of formulating musical responses to his system's playing and of listening to the music formulated by it in response to his own playing has made him "more spontaneous" and "more inventive" when, on the spur of the moment, he has to formulate musical ideas that refer to those produced by the musicians he plays with and in accordance with the harmonic constraints of a given tune.[9]

Thus, the different dimensions of the highly controlled artificial world that David had designed made this world a site in which he could cultivate the musical skills that in the jazz world are necessary for "deeply knowing another" in the real time of improvisation—skills that he was later able to apply in musical interactions with players who noticed those skills and appreciated them. What made this possible was David's decision to inform his playing with the playing of his system and thus to take a position vis-à-vis his system, which is a far cry from "the position of the 'unmoved mover'" (Helmreich 1998b:84) that many designers of microworlds occupy. In making himself an active player in the microworld he had designed, David came to occupy the position of a "moved mover," to his and his fellow musicians' delight.

This twist suggests that a more fine-grained analysis of the dynamics of human-machine interaction is needed when the implications of such an interaction are assessed against the dynamics of human-to-human interaction. Such an analysis should attend to the longitudinal dimensions of human-machine interaction and to the ways in which what seems to be the risk-free and highly controlled environment of human-machine interaction may function as a site in which individuals can cultivate the skills they need in order to function well in the risk-filled and poorly controlled environment of human-to-human interaction. Rather than jeopardizing people's ability to meaningfully interact with fellow human beings, interaction with sociable robots and digital relational artifacts may in some cases fortify it.

Trading Fours as Trading Off

At the same time, as David and I continued to train and play with his system, it also became clear that the high-level interactional context David had managed to design was an island of interactivity and context sensitivity in an ocean of the system's nonreactive and decontextualized playing, requiring David to resort to different strategies of masking, adapting to, or ignoring it. In other words, his ability to trade fours with his system turned out to represent a significant trade-off, given that his system was completely oblivious to the many forms of interactivity and context specificity that are part and parcel of the aesthetics of jazz.

In creating a meaningful jazz improvisation, musicians are expected to shape it with reference to different aspects of the improvisation's context, broadly defined. Mature players often reference what they played earlier in the improvisation, as in the case of motivic development (Schuller 1999); the contributions of other players in the band (Monson 1996; Sawyer 2003); the playing of past players, through stylistic decisions and "quoting" (Berliner

1994:195); other famous renditions of the same tune (Walser 1997); other tunes, again through quoting (Berliner 1994:368); and the broader social context through social commentary—a practice frequently and famously utilized by, for example, the bassist Charles Mingus (Monson 2007; Porter 2002). Save for referencing and reacting to David's own playing in relation to a tune's harmonic context, David's system cannot do any of these things. Indeed, it is totally unaware of these contextual dimensions. Most jazz-improvising systems lack almost all of these contextual dimensions. Some researchers have discussed this problem in relation to the general difficulty of figuring out how to imbue computerized systems with common sense (Pachet 2012:143).

Take, for example, the expectation that mature players build musical structures that gradually develop as the improvisation unfolds, rather than improvise by taking into account only one chord at a time in a tune's harmonic progression. Mature players often do so by imbuing their improvisations with a narrative-like contour, for example in the form of a beginning, a middle peak, and a conclusion *qua* denouement (Monson 1996; Schuller 1999). One of the key consequences of the architecture that David designed for his system is that every measure, and often every chord, in the system's improvisations is an island of musical activity that is disconnected from other measures and chords. The result is uniform musical activity and change, that is, musical activity and change that do not display any kind of development beyond the system's ability to reference what David had just played (cf. Manovich 2001:217).

During my fieldwork with David, we discussed this issue at length. At one point I took one of David's trumpets and improvised with the system on the tune *Four*. I took the first chorus. During the first four measures I developed a specific motif by adapting it four times to the tune's shifting harmonic context. After each of the last three times that I developed this motif, David exclaimed, "Yeah!" In so doing, he displayed his appreciation for my attempt to reach the normative ideal of motivic development. After the tune ended, I asked him whether his system can create motivic development. "No. It's boneheaded simple," he said. "If we analyzed its solos, we wouldn't discover any [motivic development]. There could serendipitously be. There could be phrases that it used, that were variations on each other and they might have happened to come out that way." "But did you not try to design it to produce motivic development?" I asked. David replied:

> No, I specifically did not, for many reasons. Because most of the computer music systems that I have heard that tried to do this sound terrible because they are heavy-handed and because they are really obvious and they're not

really interesting. This has been going on for thirty years in music research to, I would argue, ill effects. Because once you do that you're really trying to model the higher-level processes that we go through as composers, as improvisers, maybe, and that is really hard to get a handle on. And we don't really understand how we do it. It's the same reason I've never been able to come up with an automatic fitness function. The system generates surface structures. That's all. That's easy because the window is short. I can generate four bars that are going to sound great. The next four bars are going to sound OK, etc. I didn't want to train entire choruses because they would be—so I got this chorus and that chorus and they are going to cross over, and what would that mean? Nothing will come out of it.

David argues that he, and many other scholars who focus on computer music research, have not been able to develop computerized systems whose improvisations display convincing motivic development. He lumps his inability to do so in with his inability to program an aesthetic fitness function, in that both belong to "the higher-level processes that we go through as composers, as improvisers." In the context of the genetic-algorithms framework, he was unable to find a way to produce motivic development, because even if he took units or genes the length of entire choruses in an attempt to train his system to display motivic development, the eventual breeding of such long units with one another, as well as mutation, would undermine the creation of any motivic development.

Against this backdrop, David had to resort to alternative strategies to mask his system's inability to generate higher-level poetic structures, as he explained:

> One reason the system does a decent job is that it generates surface structures. The deep structure is the chord progression. It's the tune, it's the form of the tune. That's what makes it work. When it plays the blues—a twelve-bar form— you can play anything on the blues and it will sound as if you meant to. If you play *All the Things You Are* [a well-known standard tune], you can play anything and it will sound like you meant to. That's an incredibly good chord progression. It's really hard to sound lousy on this tune if you hit all the changes.

David argues that although his system's "window is short," that is, its genes are only four bars long, and although it generates surface structures, that is, notes that simply align with the scales that match different chords and in accordance with the contours of the licks in its database, the deep structure of each tune's chord progression—especially good progressions that themselves represent a form of development, such as the chord progression of *All the Things You Are*—is enough to create the impression that the system's improvisations

AN ISLAND OF INTERACTIVITY

were produced with the intention to display a higher-level poetic structure or development.

David's system not only lacks the ability to take its own unfolding improvisation as a context; it is in general entirely nonreactive to real-time contextualization cues, that is, to the signals that interactants frequently exchange with one another during an interaction and by means of which they try to point each other's attention to and/or create the context of the interaction, often with the goal of impacting its meaning and changing its course (Gumperz 1992). Jazz's highly emergent nature as an improvised form requires musicians to know how to send, interpret, and react to such cues on the spur of the moment. During our playing sessions with David's system, the system's nonreactivity to contextualization cues found expression in many ways.

For example, whereas it is easy for mature jazz musicians to signal the end of their improvised parts by means of bodily cues and how they end their parts, David had to program the trading sequences in advance each time we wanted to trade fours with his system. Thus, at one point I asked David if we could play *All the Things You Are* and if I and the system could trade parts that were initially longer than four measures and that gradually became shorter—a common trading convention in jam sessions. David responded: "Sure, how about two choruses of eights [two choruses where the system and I trade eight-bar parts] and two choruses of fours [two choruses where the system and I trade four-bar parts]? Would that work?" "Sure," I said. David then approached the computer. "All right, let's see," he said, "it's a thirty-six-bar form. Let's see how to do this." He started configuring the program to make this specific interaction sequence possible. Thus decisions that musicians would be able to make together in real time during the course of the musical interaction and in light of how it evolved had to be made in advance and could not be changed once the musical interaction began.

That David has to determine the form of interaction in advance forces him to remember it when he performs with the system, because the system cannot change its playing based on David's contextualization cues if David happens to forget the sequence or if he wants to deviate from it: "You need to remember the [preset interaction] setting," he told me before we started playing. He continued:

> If I'm in a performance and I walk around people in the audience, playing the trumpet with the wireless mic attached to it—I usually do this—and I don't remember the setting and can't look at the computer screen because I am not near it, I say to myself—"Well, I'll play a little bit and see what happens." So I'll start playing the head [the tune's melody, which musicians play before and

after the improvisations part] in case that comes up. And if it doesn't, so—"OK, I'll continue improvising." You know, you roll with the punches. Who cares? That's jazz.

Although David is right to suggest that the kind of flexibility that enables him to overcome his error or forgetfulness is a basic skill that jazz musicians are expected to master (Berliner 1994:382–383), what he leaves unsaid is that in this example his flexibility compensates not only for his own error but also for his system's inflexibility, its inability to adapt itself to David's error and thus to maintain a smooth interaction. David's system cannot engage in repair work and thus maintain the integrity of the interaction when David make mistakes (Suchman 2007). Instead, David has to perform repair work for the two of them.[10]

The system's overall nonreactivity and context insensitivity find expression in additional ways. After David told me what he does if he forgets the interaction sequence, and just before I started playing, he advised me: "Another thing is, when you play a tune with it, if you go outside the [harmonic] changes and play really weird stuff—so if you're playing with people then the piano player will follow you if he's good, if he knows what he's doing, and you'll go somewhere and have some fun, do some things, and eventually you'll come back [to the harmonic changes]. But if you play with the system and do that you'll just sound wrong because the rhythm section is going to do what it was going to do." David cautioned me to stick to the harmonic changes of the tune when I improvise on it and not to try to deviate from them, as many musicians do in order to create harmonic and melodic tension. The reason is that the system's preset rhythm section—in this case especially the piano and bass—will not adapt itself to my playing as a human rhythm section would normally do if it is attentive to the soloist's playing. Consequently, my playing would sound wrong.

The nonreactivity of the rhythm section of David's system significantly impacts one's experience of playing with the system. An example of such an impact occurred when I and the system were near the end of our joint improvisation on Miles Davis's tune *Four*. After I finished playing the melody at the end of my improvisation, I played a figurative musical ending that Davis had played in a number of his famous renditions of this tune and that has become conventionalized in the jazz world. This ending is usually played by all of the band members—rhythm section included—in a slower and much more fluid tempo than the preceding tempo. The moment I finished playing the tune's melody, I automatically played this ending, but because the system was nonreactive, it just ended its playing at the end of the melody, and I found myself playing the figure by myself. David immediately said: "That's the kind of things you cannot do with the system, the endings and all that. The system cannot change its tempo and follow you or know what you are doing when you suddenly play this ending—it's not going to happen. So I kind of end everything abruptly." Note that the system's inability to recognize my reference to Davis's famous renditions of this tune betrays another dimension of its context blindness, for, to reiterate a point I made earlier, quotations and references to famous past renditions, as well as to specific recorded improvisations of the past jazz masters, are some of the building blocks of the normative ideal of jazz improvisation.

Consider a related form of context sensitivity that mature players are expected to master and that David's system lacks. At one point David suggested that we play the Miles Davis tune *So What*. After playing the melody with David and the system, I started trading eight-bar parts with the system, and then four-bar parts. After we all improvised for a few choruses, David and I played the melody again, as is the custom in jazz performances, and I noticed that the system was now playing the melody of another tune, *Impressions*, written by John Coltrane, instead of the melody of *So What*. Because the two tunes have the same harmonic progression, it worked. Mature musicians often quote other tunes' melodies during their improvisations by adjusting them to the immediate harmonic and melodic context in which they find themselves. Such quotations are evidence not only of the musician's knowledge of the jazz repertoire, but also of his or her pragmatic flexibility, that is, the ability to use well-known musical structures in new ways and contexts and in real time, often ironically (see Gates 1988). After we finished playing, I did not comment on the fact that the system played the melody of *Impressions*. After a brief discussion about another topic, David looked at me and said: "I don't know if you noticed but on the head out [i.e., when we played the melody at the end of our improvisations] I did [i.e., program the system to play] *Impressions*—it's the same tune," he laughed. David wanted to make sure that I was aware of his decision to program his system to quote *Impressions'* melody. His demonstration of his own awareness of the pragmatic flexibility expected from mature players by means of a system that lacks it inadvertently highlighted one way in which his system falls short as a jazz improviser, compared with human musicians.

A similar dynamic occurred when I trained the system while it was improvising on the tune *All the Things You Are*. During the last two measures of the first chorus, the rhythm section suddenly stopped playing, and the synthesized saxophone, which was the instrument that David chose for the system's tone generator, improvised by itself and was then joined by the rhythm section again at the beginning of the second chorus. David burst into laughter and said, "That was a solo break," whereas I used in jest the conventional exclamation "Wow!" by means of which musicians express their admiration for an especially creative and expressive burst of improvised playing, and then I laughed, too. David's and my laughter was evoked by the fact that a solo break, during which the soloist plays alone, is typically used by the soloist as an opportunity to build emotional intensity prior to the next chorus (Berliner 1994:381). Here, too, David wanted to make sure that I understood his decision to program his system to solo during the break. And here, too, by giving expression to a structural element of jazz improvisation whose focus is context sensitivity and emotional intensity by means of a system that lacks the ability to display them, in that its playing always remains uniform in terms of its intensity, he inadvertently highlighted another way in which his system falls short as a jazz improviser compared with human musicians.

David's system suffers from other forms of pragmatic inflexibility. For example, it cannot display the kind of rhythmic variety that mature jazz players are expected to display in their improvisations. When David chose the licks from the method book he used to build the system's database of licks, he had to choose licks that consisted solely of, or that could be easily transformed into, eighth-note multiples, because only notes that are eighth-note multiples could be represented in the system's chromosome structure. Licks that did not satisfy these conditions had to be excluded. As a result, the system's ability to display rhythmic variety was significantly jeopardized. Furthermore, accomplished jazz musicians, even when they play musical phrases that appear to consist of eighth-notes multiples, do so while "swinging." The elusive notion of swinging is frequently understood as a player's ability to produce a kinesthetic response in a listener by placing notes in a specific relation to a beat. Swing encompasses subtle rhythmic nuances that are virtually impossible to represent via the standard notation system used in the method book David consulted. His system's inability to swing became a key determinant of the sensation of rhythmic stiffness that I experienced when I played with and listened to the system.

A related and equally important factor is that David's system cannot display a unique and personal sound, which is one of the key aesthetic normative ideals in jazz. At first blush, it is hard to see how developing a unique sound can be related to interactivity and context sensitivity. However, throughout the history of jazz, the ideal of developing a unique sound (and musical ideas) has been articulated by means of the ideal of self-differentiation vis-à-vis other musicians. In other words, mature players cultivate their sound in light of how other musicians sound, because having a unique and identifiable sound also means not sounding like other players (Wilf 2014a:120). David's system generates sound by means of a simple tone generator that can simulate the sound of different instruments. The sound is heavily synthesized and is not a great approximation of the acoustic instruments it supposedly simulates. Most important, it sounds as standardized as any other mass-produced sound-generating electronic technology.[11] As I noted earlier, according to David, because his system "can play any instrument that I choose" and thus affords him a high level of control, he is not bothered by this issue. His aesthetic choice, however, violates a key normative ideal in the cultural order of jazz, whose historical roots reach to what one scholar called "the heterogeneous sound ideal" prevalent in many African and African American musical traditions (Wilson 1999:160).

The standardized quality and nonexpressivity of the sound of David's system have concrete implications beyond the abstract violation of a key aesthetic ideal. In the jazz cultural order, the emblem of a unique sound and form of expression is the human voice. In the same way that each person has an identifiable voice, players should strive to develop a unique sound with their instruments. Furthermore, they should attempt to attain with their instruments the same high level of expressivity and immediacy of execution of ideas that the voice is thought to afford singers (Wilf 2014a:191). Given that the human voice is considered to be an emblem of uniqueness of sound and of human expression and that it is diametrically opposed to what the synthesized sound of David's system represents, David decided to avoid simulating the human voice with his system, despite proclaiming that he is "more interested in the notes that [the system] plays and that they are executed correctly" than in their sound quality. During our playing session, he told me about a trio that he used to play with, where "the guitar player who sings has a nice kind of vocal quality. And we did some tunes that were fun to play. And there's no way I could do those with my system. I can't do vocals with it. If it did vocals—I don't even want to go there! It would be a disaster."

Against this backdrop, David resorted to different strategies in an attempt to make the notes generated by his system sound more human-generated. He explained:

> I have a few heuristics to make it sound more like a jazz player. This is really boneheaded simple. Basically, you know how MIDI works: it's very precise. So what I basically do is, I go in and find a parameter I can tweak and randomize and then set a slide runner, if you will, a constant, where it will randomize it within a range, and jack up that constant until it gets annoying. [The parameters are] no precise onsets, a little early or a little late. Note length—a little longer or a little shorter. Loudness—eighth notes off the beat get hit a little harder

than notes on the beat—but it's all kind of randomized. Pitch—every pitch is hit a little off-kilter, and then if the pitch is held it asymptotically approaches the correct pitch. The deviation is half each time. I played with the envelope generator on the tone synthesizer. Most wind instrument players, when they hit a note, they hit it a little flat and then they slide into it, so I'm imposing that kind of pitch envelope on everything it does. This is why I tend to stay with wind instruments. When you do that on a vibraphone it sounds weird. Vibes can't do that—they can vibrato, but they can't pitch bend. So onset, the length, and the loudness—how hard it gets hit, and a bunch of others. . . . I never want it to sound like a program. So if you turn off all those factors, it sounds pretty wooden. You turn them all on, and it does a reasonable job.

Despite these attempts to make his system's playing less "wooden," the system's rhythmic, expressive, and overall interactive inflexibility and context insensitivity significantly limit the kind of music that David can play with it. For example, David tends to play only medium- or up-tempo tunes. If he decides to play slow tunes, he must resort to different strategies to mask his system's inability to function well in that musical context. "On fast-tempo tunes it usually sounds way better," David told me after we finished playing *Four* in a quick tempo. "That's one of the reasons *Giant Steps* works so well.[12] When it's playing at 240 [beats per minute] you don't really hear all the stuff that it's doing. So it's like—'Wow.'" "Can we now do a ballad?" I asked. David said: "I don't do true ballads. I don't really do anything that is below 120. So when I do something that is slow, I will do the solos in double time. So that's what you have to work with." Playing ballads typically requires musicians to use and sustain longer notes. Ballads thus focus the audience's attention on the quality and expressivity of the soloist's sound more than fast-tempo tunes. In addition, slow tunes open up more space between the beats. In order to fill this space, the soloist must display great rhythmic flexibility, as well as be more in tune with the rhythm section. Because David's system cannot display all these features, he designed his system to play double time in the few slow tunes that he does play with his system. This means that the system will continue to play fast successions of short notes on these slow tunes as well.[13]

An Abstract Refuge from an Abstract World

Against the backdrop of the kind of frustration that had motivated David to embark on his long journey, the limitations of the solution that he ended up with at journey's end suggest that he inadvertently reproduced by means of his system some of the core dimensions of the social reality that had been responsible for his frustration. This surprising result points to the inescapable social embeddedness of digital systems, even, and perhaps especially, when those systems are designed as a kind of escape from social reality.

To begin with, consider the fact that during our conversations David frequently attributed his unsatisfying experience of playing in jam sessions to the ill effects of academic jazz education:

> Some of the jazz students I've played with, I would pick my system over them any day because I know what practice book they saw. I mean, they are working on their technique, they're students, they are not finished products, I understand. They are a work in progress, and some of them are monsters in terms of being able to operate their horn and just execute things. But in terms of their ideas, there aren't many. There are just a lot of notes. My system will generate more interesting solos, to my ears, anyway, than standard ideas that you've heard fifteen times. It gets back to how you come up with ideas and all that. And if you look at a lot of the improvisation books, they're telling you to do this pattern over this chord chain so that you'll hit the leading tones on the third or the fifth [degrees of the chord] or whatever, which is a theoretical idea, but that is what you should do to improvise. And most jazz kids coming out of high school or these academic programs, that's what they have been taught because they have been taught by people who know this theory and believe in it and that's how they think you should play.

"But your system takes into consideration chords and scales too," I said, somewhat perplexed. David responded:

> Yes, it does. It's diatonic. There are things that result from jazz theory, like the chord-scale mappings and all those things that are pretty clear, and you can look at it and say that it validates a lot of those method books that say, "Gee, if you have a C minor seventh chord, you ought to play this scale." It does this, sure, but it adds something different. This goes back to the mutation operators and what it's doing with them. And the fact that it's going to get it right the first time every time because it cannot—it cannot play a theoretically wrong note—so it's that and also that any melodic contour that it works with is always going to be mapped correctly to the harmonic changes and it will play notes that are reasonable.

In itself, the fact that David articulates his displeasure with jazz musicians in terms of the impact that academic jazz education has had on the American jazz scene is not surprising, because most contemporary jazz musicians are now likely to receive at least some of their training in academic programs. What is noteworthy is that David's system reproduces many of the ill effects he attributes to the rise of academic jazz education, and others that have been attributed to the rise of academic jazz education not only by outside critics and professional musicians, but also by academic programs' administrators, teachers, and students.

Consider the fact that David faults jazz schools for teaching students to improvise based on chord-scale theory as it is codified in mass-produced method books. Chord-scale theory is supposed to help players improvise on a given tune by spelling out for them the relationship between the chords that comprise the harmonic basis of the tune and the scales that correspond to them. Thus, players can infer which notes they can use in constructing their improvisations. This body of expert knowledge was initially developed by different musicians and educators in the jazz world around the mid-twentieth century (Ake 2002; Monson 1998; Owens 2002). As jazz training became increasingly institutionalized in academia, this theory became more systematically codified and widely disseminated through different pedagogical aids—mainly method books. Critics of academic jazz education have targeted chord-scale theory as a key factor in the detrimental effects that academic jazz programs have had on their students' improvisational skills (Nicholson 2005). They have argued that because students are taught to rely on chord-scale theory, they learn how to improvise by practicing patterns and licks that they then apply to the harmonic changes, while making their point of improvisational reference the single chord. Thus, when improvising on a harmonic progression, students typically progress from one chord to the next, and their improvisatory considerations are reduced to playing the "correct" notes on each chord, one after the other. Students' improvisational solos consequently become a long succession of similar patterns of improvisation that bear no relation to the tune's melody or broader form and that display no poetic structure or development that exceed the limits of one chord at a time within the tune's harmonic progression (Wilf 2014a:144–146). Against this backdrop, it is noteworthy that David's system not only relies heavily on chord-scale theory, but also produces improvisations characterized by low-level musical meaning that exists mostly on the level of half a measure or one chord, except when the system trades fours with its human interlocutor, as David himself acknowledged during our conversations.

David's system reproduces some of the other negative consequences that academic jazz education has had on the proficiency of many contemporary jazz musicians and that can be directly and indirectly related to what David perceives as the unsatisfying playing of his fellow musicians in the jam sessions he had attended. The predominance of chord-scale theory in academic jazz programs can be attributed to the fact that such programs are predicated on the rationalization, standardization, abstraction, and codification of rule-based knowledge in textbooks (Wilf 2014a:119–120). According to critics, this bias toward literacy-based modes of knowledge transmission and reception at the expense of teaching and learning via real-time playing in live settings has impaired students' ability not only to display motivic development in their improvisations, but also to respond to the playing of their bandmates in real time (202–208),[14] to develop rhythmic variety beyond playing rapid successions of eighth notes (144), to swing (120–121), and to cultivate a unique sound and project emotional and expressive intensity (191–192)—precisely the same dimensions of interactivity and context sensitivity that David's system lacks, as I explored earlier in this chapter.[15]

How could David have ended up designing a system that reproduces many of the problems that are responsible for what he considers to be his fellow musicians' flawed playing? One reason may be that both the root of the problem that David had attempted to escape—academic jazz education—and the solution that he devised to this end—his digital system—are structured to a significant extent by the same epistemological logic of rationalization, standardization, abstraction, and codification of knowledge in some kind of symbolic language, whether it is a music-notation system or a programming language. This epistemological logic excludes the myriad aesthetic dimensions of jazz as a collectively improvised and aurally mediated form of situated musical creative practice.[16] One potential lesson that can be derived from this consequence is that because the real or imagined insufficiencies of human interaction are embedded in specific social and cultural contexts, the solutions to those insufficiencies are best developed in relation to those contexts rather than away from them.

Withdrawing from Humans, Withdrawn from by Humans

A second unexpected consequence of David's adventure with computational creativity is that his withdrawal from human musicians as a solution to his frustration with them has led to humans' withdrawal from him, which he experienced as a new kind of problem. First, the price of replacing human musicians with a computerized system has been the constant risk of audience indifference whenever he performed with his system in live settings. David explained to me:

> From an audience perspective, they need to see the interaction, they need to see that there are different things changing. This is why I don't take long solos and the system doesn't take long solos because given the visual thing, what am I going to do when it solos? Go and buy a drink? All that would be left is a little computer on a table. The audience wouldn't know what to do with that. When

> I played with people, you know, I would play the head [melody], perhaps take a solo, and I'm done for five minutes while everybody else solos for a while. But with the system I'm done for one chorus and then it solos but then we're trading fours or eights—I'm playing a lot! I have to be playing a lot!

David argues that he cannot leave his system to improvise by itself for too long because the audience would quickly get bored, since all they would see is a laptop computer on a table (see fig. 2.1). Therefore he has to play a lot during his performances with the system.

In addition, the risk of audience indifference has made it difficult, if not impossible, for him to secure gigs in commercial performance venues:

> DAVID: The bad news is that nobody wants to book it because they think it's a CD or something.
> EITAN: Even if you explain to the club owners what it is?
> DAVID: Oh yeah. You explain it but it means nothing to them. What does "real time" mean? That's a technical term. Any technical term—what's a genetic algorithm? What's evolutionary computation? What's bidding? What's a real-time interactive performance system? If I'm in an academic setting, people know what it means and all that. But club owners are not academics. They're really not academics. They're businesspeople with a room to fill, and they are trying to figure it out.

Owing to the risk of audience indifference and the difficulty of securing gigs, David's performances with his system have been limited mainly to official functions at his university and to academic conferences.

David's long journey to replace musicians with a machine also coincided with the decision made by some of the musicians he had played with to replace him with a vocalist, as he explained:

> I haven't played in a stable group for a long time, with people. I mean, there's Alex, a piano player. We've been in this jam session together for seven years and we used to play anything from a duo to a quintet. And we played coffee houses as a duo, Alex on piano and I on an amplified trumpet, although it was usually quiet enough to be open [i.e., not to use amplification]. And it's a lot of fun. We did some blues tunes. I miss that to some extent. When I first started playing with my system I played a couple of gigs with Alex and I unplugged the piano player from the system's rhythm section and I had Alex play the piano. It was really disorienting for everyone. It didn't work. It just didn't work. And about that time he started to use a vocalist instead of me with his trio, and of course if you use a vocalist you will get hired. Vocals are what people are looking for in clubs because for most people music is lyrics and without lyrics

it doesn't exist. It's just this thing out there [for them]. So he went off to do this thing with the vocalist for several years and by that time I was with my system and so, it didn't happen that way on purpose, but we just kind of went on our separate ways. And I run into him every now and then and I sat in with him a few times through the years and I kind of miss playing with him.

Recall that in the cultural order of jazz, the human voice epitomizes the normative ideal of developing a personal sound with one's instrument and of projecting emotional and expressive intensity with it during one's improvisations (Wilf 2014a:191). It represents the antithesis of what David's system is capable of doing. For this reason, David decided that the human voice would be the one instrument that his system would not simulate. Against this backdrop, the decision made by the musicians with whom he used to play to replace him with a vocalist represents an unexpected and somewhat ironic turn of events, in that this decision is diametrically opposed to the one made by David to replace these musicians with a digital system. More important, this turn of events has made David realize that he missed playing with, not only for, people, as well as to reassess the value of playing with them versus with his system. This kind of reappreciation for human creativity and sociality following attempts to simulate, enrich, transform, or escape them by means of digital computation is not unique to David's case. As the next two chapters demonstrate, it characterizes another project focused on building a jazz-improvising system—one that is markedly different from David's system in terms of its goals and design.

3

"A Device That Would Generate New Musical Ideas":
The Computer as a Source of Musical Inspiration

"They All Sound the Same": Fixity of Style and Its Discontents

I first met James, one of the directors of the lab in which I conducted fieldwork, on the eve of a concert in which Syrus and other robotic projects that James's team had developed were to be featured. Syrus, a five-foot-tall robot marimba player, was the result of a research project on robotic musicianship, funded by the National Science Foundation, that James had initiated at the institute of technology where he was a faculty member. I had sent James an email a few days earlier in which I explained my interest in conducting fieldwork in the lab. Specifically, I was intrigued by James's ongoing research on the design of robots that improvise rather than play precomposed pieces. In my email, I described my previous study of contemporary modes of socialization into jazz in academic programs in the United States as a way of framing my interest in Syrus. I explained that my previous research had focused on the rationalization of jazz training in higher education, and that James's attempt to develop a jazz-improvising digital computational architecture that could animate Syrus's playing might be conceptualized as an extension of this rationalization. However, it soon became clear that James had a totally different interpretation of what he was trying to achieve in relation to academic jazz education.

    I arrived at the concert hall a few hours before the beginning of the concert during what seemed to be a break in a rehearsal. Syrus was positioned onstage in front of the marimba. James and his students were sitting amid open pizza boxes and talking to each other. After introducing myself to him, I explained again my previous research on academic jazz education. I was in the middle of a sentence when James interrupted me: "They all sound the same." At first I was confused. "Who?" I asked. James replied, "The students! They all sound the same. Like machines!" He laughed and continued: "And all the musicians who come out of the schools, and like 99 percent of the jazz musicians today—they all sound the same. You know what they say: 'Jazz may not be dead but it sure smells funny.'" As I was thinking how to respond, James went on: "This is why I built Syrus. Because I wanted to be inspired. I wasn't inspired anymore by—everything that can be written had already been written. Everything that can be played had already been played. I felt that I understood all the genres that I was familiar with, like jazz—there was nothing that really caught my interest, a new sound, new ideas. I wanted to develop a device that would generate new musical ideas that I could not come up with by myself, nor could other people." James's explanation brings to mind David's discontent with the musicians he had played with, which led him to develop his system. James argues that jazz musicians stopped inspiring him. When he listened to the music that they played, he could not hear anything new and exciting, only the same old musical ideas. His disappointment led him to assemble a research team in order to develop Syrus. Like David, James articulates his critique of jazz musicians by means of a critique of academic jazz programs and their graduates. As I mentioned in the previous chapter, this critique is voiced not only by outsiders like James and David but also by these programs' administrators, teachers, and students. James argues that, given that such programs function as the primary site of jazz socialization in the contemporary United States, "99 percent of the jazz musicians today . . . sound the same."

To be sure, James's explanation seems contradictory: he appears to argue that he wanted to build a machine that would inspire people because jazz musicians have become "like machines." However, the two kinds of machines that he is talking about are completely different. Critics of academic jazz programs and the standardizing impact that they have on the playing of their students often reference the machines that are associated with the Industrial Revolution. Such machines are based on repetitious and preprogrammed action along the model of the assembly line and are associated with the mass production of iconic standardized consumer goods such as cars (Wilf 2010:567–568). In contrast, Syrus, like David's system, is a machine of a different kind. It is based on a digital computational architecture that integrates stochastic processes into its logic and whose output is consequently seldom repetitive or predictable. It is a machine that is supposed to simulate the desired contingency of human creative action and thus to help players who have come to sound like the machines of yesteryear to enrich their creativity.[1]

However, as became clear on the following day when I met James for lunch, his criticism of the sterility of contemporary music did not take its inspiration solely from the machines that characterized the Industrial Revolution. It was also directed at the way in which the same digital computational architecture that is at the center of his research has been used in the broader field of algorithmic music composition. For example, the computerized algorithms known as Markov processes that animate Syrus have been used in algorithmic music composition ever since the 1950s (Nierhaus 2010). They are especially suitable for style imitation based on the analysis of large corpora of music. James took issue with scientists and composers who use these algorithmic processes to simulate well-known and already familiar musical styles. At one point during our conversation, James mentioned The Continuator, a software developed by the music technology scientist François Pachet (2003), whose work I brought up in earlier chapters. This program is designed to learn in real time the improvisation style of a player while the player is improvising and then continue the player's improvisation in the player's style when the player stops playing. James said:

> I think that The Continuator is more successful than us [his research team] in capturing a given style. It does more complex things than we do. But, on the other hand, I personally think that it was less successful than Syrus in inspiring because all it does—which is an achievement, don't get me wrong—is to capture—"Wow, it sounds like Chick Corea playing," or "Wow, it sounds like me."[2] When I played with The Continuator I said: "Yes, this is my style," but it did not inspire me because I already knew how my style sounds, you know? The Continuator and similar systems are supposed to impress you by playing different things but in your style. It's not supposed to produce a different style. And that's not enough for me. There's nothing inspiring in that for me. And it's not a problem to produce a system that creates music that is not interesting.

James's critique focuses not only on repetitious machines and musicians but also on machines that integrate stochastic processes into their computational logic to mediate existing styles. Such machines do not inspire him.[3]

David's system represents precisely the kind of jazz-producing digital computational systems that James criticizes. David was content with developing a system that would play in his own style because he was more bothered by the lack of conversation and interaction between players in the real time of performance than by the fact that players played in styles that have existed for over sixty years. Consequently, his goal was to design a system that would play in his style and that could tightly listen and respond to his own playing, and to whose playing he could tightly listen and respond—all in the real time of improvisation. For James, in contrast, the more acute problem was the fact that jazz players have been playing in the same familiar styles for many years, to the point where they all sounded alike. Consequently, he decided to design a machine that would create new improvisation styles with which jazz players could interact and that could inspire and influence their playing.

Indeed, the more I listened to James's description of his goals, the more they appeared to be the exact opposite of David's. During our lunch, James described a conceptual shift that he had experienced at the early stages of his research:

> My purpose is not to build a tool for you as a musician, but rather a computer that is a musician in itself. It's the difference between the computer as only an instrument for you to use and play with, and the computer as a musician that activates you and that can really inspire you. So I made a switch from "The computer is here to serve you" to "The computer is here to inspire and challenge you." And once you want to have that you need to give the computer everything that a musician has—embodiment, social cues, visual cues, and the acoustic sound—very important! The worst thing that can happen is if people will be satisfied with robotic systems that reproduce what you already have, using the same kind of tricks and shticks. It will be a Berkleeization all over again, only this time of a robotic kind. You don't need to develop another system that plays in the same way that musicians already play today, but a machine that can invent new styles, like Miles Davis did throughout his career.

A number of points are worth noting about James's description of his goals. First, James characterizes machines that simulate existing styles, such as David's system, as quasi-subservient instruments that do not alter or challenge what human musicians already know and what they can already do, as opposed to machines that are designed to function as independent musicians that can challenge and inspire them. On the surface, his argument resonates with the scholarly framework that I discussed in the previous chapter, which distinguishes between "the computer as instrument paradigm" and "live algorithms as real-life partners" (Young and Blackwell 2016). However, his critique of systems such as David's actually qualifies this scholarly framework that stipulates that the ability to "generate output independently of performer and designer" and to "offer demonstrable autonomy with a capacity to both react and contribute constructively" (511) is a feature of "live algorithms as real-life partners" rather than of "the computer as instrument." Although David's system is able to display this feature, James suggests that because systems like it are designed merely to mirror a musician's existing style, they are incapable of truly displaying the kind of autonomy that can challenge musicians and inspire them to explore new creative worlds. James's critique also qualifies the suggestion that one criterion that can differentiate between the computer as an instrument and the computer as "an independent improviser" is whether or not "the system is designed to avoid the kind of uniformity where the same kind of input routinely leads to the same result" (Lewis 2000:36). David's system will never produce the same result given the same kind of input, yet James would still consider it more of an instrument, because its goal is to mirror David's style of playing rather than to challenge it. James's critique suggests that a more fine-grained typology of artificial music systems is needed based not merely on their technological capacities, but also on how they are put to use, to what ends, and in what historically and culturally specific contexts.

Second, James cautions that designing machines whose only goal is to mediate existing styles will simply reproduce in a digital, computerized form the negative consequences of contemporary academic jazz education and its associated reactive, conservative, and standardizing biases, which he invokes by means of this form of education's most iconic institution, Berklee College of Music (Wilf 2014a:15). Indeed, as we saw in the previous chapter, David's system has ended up reproducing a number of the shortcomings that characterize the playing of academic programs' students, except for the ability to engage in a real-time musical exchange with one another. James argues that machines capable of truly inspiring and engaging musicians rather than merely entertaining them and mirroring what they already know must be endowed with a complex embodied form that can provide them the basis for exchanging "social" and "visual cues" with their human interlocutors and for producing a meaningful "acoustic sound"—precisely those dimensions that are often neglected in the jazz academic setting. David, in contrast, was perfectly content with the fact that his system plays in his style and affords him complete control over many dimensions of the performance situation. Keeping his system's computational architecture and embodied form as simple as possible was a condition of possibility for providing and maintaining such control.

Last, note that both David and James invoke the legendary trumpet player Miles Davis in their descriptions of their respective goals. However, each invokes a different aspect of Davis's career and creative persona. Recall that David told me that his desire is to sound like Miles Davis at the early stages of his career, when Davis played lyrically and inside the harmonic changes, and that he designed his system's playing to align with this style. In contrast, James focuses on another dimension of Davis's career for which he became well-known: his uncanny ability to reinvent himself stylistically (and, consequently, to change the history of jazz) a number of times during his career.

To sum up, although James's disappointment with other players partly resonates with David's disappointment with them in that both experienced them as predictable and uninspiring, each focused on a different dimension of the problem. Consequently, each developed a different solution by means of a different kind of jazz-improvising digital computational system. Whereas David was content with designing a machine that would improvise in his own style, James hoped that his system would help him and others to explore and create ideas in new styles—"ideas that I could not come up with by myself, nor could other people."

On the surface, the differences between David's and James's goals can be conceptualized by means of the typology of kinds of creativity suggested by Margaret Boden, as I discussed in the introduction. David wanted to design a machine that could perform the task of "exploring [already existing] conceptual spaces" or "structured styles of thought" (Boden 2003:4). Such conceptual spaces are spaces of possibilities that derive from a given set of constraints. Within this form of creativity, any novelty or "a new trick" is "something that 'fits'... [an] established style: the potential was always there" (5). In contrast, James wanted to design a machine that could create new conceptual spaces or styles, "so that thoughts are now possible which previously... were literally inconceivable" (6).

However, a close analysis of the ways in which James and his team members tried to achieve their goals suggests that these types of creativity do not correspond to clear-cut formal properties that can be identified in the operation of this or that system. Rather, the difference between these two types of creativity is performatively created and socially constructed in the course of the design and use of the systems that are supposed to exhibit them.

Mixing like a Machine: Computational Alchemy's Allure

To build a robotic improviser that could play with human musicians and inspire them to explore new styles, James decided to build a robot that would "listen like a human and play like a machine," as he told me. Accordingly, his research team focused on two types of algorithms: perceptual and generative. Perceptual algorithms enable Syrus to "listen" to the music produced by the human musicians who interact with it by means of a digital interface, and to detect in this music a number of meaningful dimensions. At the time of my fieldwork, the algorithms processed a note's pitch, duration, and, to a lesser degree, volume. The research team was in the process of designing the system to perceive harmonic tension—the relation of notes to the harmonic context in which they are played. These perceptual algorithms are in charge of the "listening like a human" part.

The role of the generative or improvising algorithms is to generate the music that Syrus would eventually play in response to the meaningful dimensions that it detects in the music played by its human interlocutors. Syrus's main algorithmic architecture is based on hidden Markov models, which integrate stochastic processes into their logic and whose output is seldom repetitious and predictable. James decided that, rather than creating a database of standard musical phrases or licks for his system to process, Syrus's databases would consist of recorded past improvisations of well-known jazz masters. For each jazz master in whose style they wanted Syrus to be able to play (with the ultimate goal for Syrus to later mix those styles and thus create new ones), the members of the research team created a large database of this master's solos. These solos are in a MIDI (musical instrument digital interface) format, which means that the files can be fed into a computer program that can break the musical information into chains of pitch and duration data, which are represented numerically. These data are then analyzed against chord-change score files (the harmonic sequences on which the player improvised). The system statistically analyzes this corpus to generate transition probabilities, that is, the probability that a certain future state will follow a given present state. During performance, and for each note played by Syrus, the digital system constantly searches for a match between the last sequence of notes performed by the player who plays with Syrus (if Syrus takes turns with a player) or by Syrus itself (if Syrus improvises by itself) and the chains of pitch and duration values derived from the jazz master's corpus, which are stored in the system's memory. The length of the sequence is usually determined in advance (two or three notes each time, for example—a point I will elaborate on later in this chapter). Any such search yields a number of candidates. The system chooses stochastically—that is, based on chance decisions weighted by a function of likelihood, itself determined by the statistical analysis. When a matched sequence is selected from the system's database of the master's solos, the system instructs Syrus to play the note that continues this sequence as it appears in the memory—that is, to play the note that the master had played after he or she played that specific sequence. The system's decisions (the notes Syrus plays) feed back in real time as new input, and thus the decision process begins again. All this computation takes place in a split second prior to every note that Syrus plays.

One of the main reasons James decided that the system's database would consist of representations of the recorded solos of jazz masters is that he had hoped to create new styles of improvisation by mixing the masters' existing styles as synthesized by the system from the databases of their solos. As James told me, "What we did [in the lab] was perhaps less sophisticated statistically [than The Continuator]. But combining and morphing different styles—people have not done this before. This is our novelty. In this way we can generate responses that Chick Corea would have never thought of because

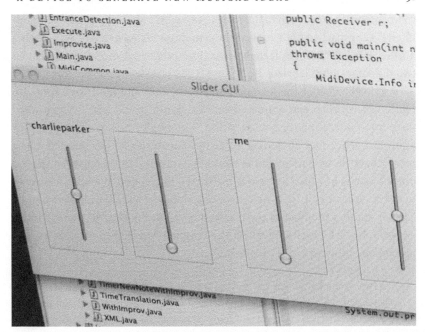

FIGURE 3.1. Sliders for controlling proportions of different styles in Syrus's playing.

suddenly it's 60 percent Chick Corea and 20 percent Miles Davis and 20 percent you. This is where I expected the inspiration to come from, which you cannot get from humans." Team members can set and change the proportions of such different styles in Syrus's playing via sliders on the software interface (see fig. 3.1).[4]

As I soon learned, the research team did not limit itself to experimenting with different mixes of past jazz masters' improvisation styles. One day, as I approached the lab, I heard a strange mix of sounds coming from behind the closed doors. The sounds were both familiar and strange. I opened the door with my access card and entered the room. Kim, a graduate student who was a member of the research team, was playing on the electric keyboard that was plugged into the computer controlling Syrus. He looked intently at Syrus while playing. Matt, another member of the research team, was looking at one of the computer monitors. I recognized the first movement of Beethoven's "Moonlight" Sonata, which Kim played with ease. Suddenly Syrus started to play on the marimba while Kim continued to play the sonata. I listened carefully. I could identify in Syrus's playing rhythmic and melodic motifs taken from what Kim had just played, but these were only hints. Syrus did not copy Kim's playing but rather loosely wove bits and pieces from it into its own playing. At one point Kim turned to Matt and said, "The jazz style produces much more interesting results than the rock style, right?" Matt answered, "Yes, definitely!" I approached Matt, looked at the monitor and saw that he was shifting a slider labeled "Jazz" with the cursor. Two other minimized windows had two different sliders labeled "Rock" and "Classical."

As I realized after this session had ended, Kim and Matt were trying to figure out which preprogrammed musical genre would lead Syrus to produce more interesting responses to the "Moonlight" Sonata. These preprogrammed genres were based on a statistical analysis of a few selected music pieces that belong to them. In the weeks that followed this session, I observed other sessions in which Syrus was instructed to respond in a "classical style" to a jazz piece or in a "jazz style" to a classical piece played or improvised by Kim or another member of the research team on the electric piano. I interviewed John, a doctoral student who had originally compiled these styles, and he explained this aspect of the research:

> That was something I was just playing around with. What I was interested in was what if I take, for example, Mozart and Chopin and mix them and then introduce it to a jazz standard and see what comes out. Or another time I was going to throw a couple of Romantic composers together and then I was going to throw modern harmonic language at it and see how it responds. With all those styles, I was just curious what would come out because we had this engine there and it can treat all the music the same way, so I was just curious what would come out.... It definitely created interesting results that were a lot of fun to play with. And it felt new. It wasn't something you heard before. That was the part that I found the most interesting in the research.[5]

The research team's modes of experimentation, as well as John's and James's descriptions of such experimentation, point to the nature of the pleasure that team members derived from the prospect of mixing improvisational and compositional styles, as well as entire musical genres. Their pleasure resulted not only from their feeling that a style or a musical genre, which are commonly understood to be abstract entities, could suddenly be made concrete in the form of lines of code that can be executed with a stroke of a key, but also from their feeling that different styles and genres could suddenly be mixed with one another with the same scientific precision with which a chemist mixes different substances to achieve specific results.

On the one hand, such experiments resonate with the fact that style-mixing has provided the basis for creative developments in different forms of art, from jazz (as when jazz musicians such as Dizzy Gillespie borrowed elements from Latin American music) to abstract modern art (as when painters such as Matisse and Picasso borrowed from African sculpture). On the other hand, they are based on the reduction of entire styles to the statistical distribution of a very limited number of dimensions, such as pitch and duration. Such experiments consequently resemble only on the surface the highly complex nature of style-borrowing in art. In practice, they depend on the radical simplification of styles and of the notion of style-mixing as a condition of possibility for their alchemic promise and for the excitement they consequently generate. Recall the vignette with which I opened this book, in which one of the students who sat next to me in the lab asked his two friends, "How much 'Miles' will you have in your cocktail, sir?"

The excitement that team members experienced can be further accounted for against the backdrop of linguistic anthropological approaches to the analysis of style as a dynamic feature in real-time communicative events. Such approaches have conceptualized individuals not only as possessing sets of distinct communicative styles, but also as possessing the capacity to generate such styles in real time in different complicated ways (cf. Eckert and Rickford 2002; Goffman 1981; Bakhtin 1982). However, if "all linguistic patterns of use arise from decisions people make in interaction when they are talking to a real person and thinking about 'who they are' in relation to that person or people" (Kiesling 2009:172), "these decisions are not necessarily 'conscious' in the sense of being open to reflection, in the same way that we do not calculate all of the actions necessary and do calculus in order to catch a ball" (192). The excitement experienced by team members resulted from their feeling that they can make style-mixing an object of heightened conscious agency by delegating "the calculation" to a nonhuman agent (Callon and Law 2005), that is, the digital computer, which they could manipulate at will (Hayles 2009:38). When John explained to me matter-of-factly his and the other students' experimentation with mixing different styles by saying that "we had this engine," he revealed how the access to such a nonhuman agent can naturalize the fantasy of bringing intentional agency and real-time style manipulation under one roof as never before, based on ideas that reductively conceptualize creative agency (and agency in general) as a series of calculations.

"Stylistic Fidelity": Reproduction as Creation of Styles

Inasmuch as the promise of mixing styles, which is at the center of James's research, depends first on making styles concrete and representable as digital files, the critical analysis of this promise should first focus on the series of operations that support this initial stage. The critique of the notion of sound fidelity that scholars have made in the context of technologies of sound reproduction can productively inform the critical analysis of media technologies that promise to deliver what I call "stylistic fidelity" (Wilf 2013a).

Jonathan Sterne has argued that the emergence of sound-reproduction technologies created the fiction of sound fidelity, that is, the idea that such technologies can produce copies of original sound events, which are acoustically identical to them, as if such sound events existed independently of and prior to their recording. Based on an extensive historical analysis of such technologies, Sterne shows that original sound events were in fact the product of a radical "configuration of bodies and sounds in space, a particular ordering of practices and attitudes" (Sterne 2002:236). Recorded sound became studio art, an orchestrated production of sound events arranged specifically for the purpose of their recording, during which singers and performers strapped to microphones needed to learn new performance skills. The very technology of sound reproduction thus created a new form of sound event: "The very idea that a reproduced sound could be faithful to an original sound was an artifact of the culture and history of sound reproduction. Copies would not exist without reproduction, *but neither would their originals*" (282). Drawing on Sterne's analysis, I suggest that what at first appears to be the reproduction or mediation of well-known jazz masters' improvisation styles that have a prior or independent existence is, in fact, the reproduction or mediation of new styles that have been created specifically by team members for the purpose of their successful reproduction or mediation and that would not have otherwise existed.

To begin, team members were highly selective with respect to the jazz masters whose styles they chose to reproduce. When I asked Randy, a graduate student who was in charge of the research project's style-mixing dimension, why the team chose to focus on players such as Miles Davis, Thelonious Monk, and John Coltrane, he responded:

> Well, that was part of finding two or three jazz performers who are highly noticeable, that we know what their styles are. I would say they are high profile in the sense that it's much more likely that the average person will know something about their style. And also their styles are distinguishable from each other. Those are the two factors that have led us there. . . . I can tell you—"Oh, this is clearly a Miles Davis solo," or "This is clearly a Coltrane solo"—you know. They are more noticeable. So these were the two factors that went into that.

Randy explains that team members did not select just any jazz master but only those whose improvisation styles could be easily recognized and differentiated by listeners.

Furthermore, once team members chose a specific jazz master for style reproduction, they performed a similar kind of selection for recognizability and differentiation purposes on the level of this master's corpus of recorded improvisations. They never created a database that consisted of the entire recorded oeuvre of a past jazz master. Rather, they intentionally chose to feed their technologies recorded improvisations from a stylistically identifiable period in a master's career and, moreover, to leave out specific recordings made in this period that might jeopardize the identifiability of his style (cf. Manovich 2001:224–225). When I asked Randy to describe how the team selected a jazz master's corpus to function as a database, as well as the typical size of such a corpus, he told me:

> We found that it varies between author and author, how much we need to get a sense of their style. But at the very minimum I would say four or five pieces. And, of course, across a person's lifetime their style changes dramatically. So if we are trying to get early Coltrane, we had to focus on grabbing pieces from that period. Because obviously in a lifetime the style changes a lot and it becomes less meaningful how close the representation is if you just throw everything into the same style. And the interesting aspects of this are—you have to be careful with what you choose because there is a lot of weighting that goes into it when you're dealing with low numbers like three or four pieces.

Team members' decision to choose easily recognizable styles for reproduction brings to mind Sterne's argument that during the development of sound-reproduction technologies, their inventors chose to reproduce sonic phenomena that were easy to discern and that the wider public was likely to be familiar with and thus to recognize, despite their noisy and messy reproduction: "In the process of creating and testing a machine designed to reproduce sound as such, certain types of sound were privileged as ideal testing material—specifically, easily recognizable forms of human speech. This kind of speech was limited and particularly conducive to reproduction, that is, easily understood by a listener with relatively few explicit cues to go on: rhymes, popular quotations, newspaper headlines. . . . In other words, conventionalized language helped the machine along in doing its job of reproducing. It enacted the *possibility* of reproduction before that function could be fully delegated to the machine" (Sterne 2002:247). Sterne notes that Alexander Graham Bell was an elocutionist, someone whose speech was already fit for reproduction by means of the telephone he invented and whose functionality he demonstrated by speaking into it (Sterne 2002:247). Similarly, "when Edison's lab was able to construct a fully functional phonograph . . . , Edison's famous test quote was again language easily remembered and easily understood": the song "Mary Had a Little Lamb," which was not only well-known but also had a number of redundancies built into it (such as rhyme and rhythm) that aided its recognizability despite its noisy reproduction (250).

Finally, after team members chose specific recorded improvisations from a specific career stage of a specific jazz master, they performed a similar kind of selection on the level of the bits and pieces of each of this master's chosen recorded improvisations. To recall, Syrus's algorithmic architecture is based mainly on Markov processes, which constantly search for a match between the last sequence of notes performed by the player who plays with Syrus (if Syrus takes turns with a player) or by Syrus itself (if Syrus improvises by itself) and the chains of pitch and duration values that were derived from the analysis of a jazz master's recorded corpus and that are stored in the system's memory. When a matched sequence is selected from the system's database of a master's recorded solos, the system instructs Syrus to play the note that continues this sequence as it appears in the database. Crucially, team members needed to decide in advance on the length of the sequences of notes that the system would search for in its database of a master's recorded solos in response to the music played, which the system would continue "in the master's style." Such a decision, too, in fact produces the style that appears to have an independent existence prior to Syrus's mediation of it. A higher-order sequence (such as a specific sequence of five notes) would result in fewer candidates, because it was likely to occur fewer times and be followed by the same notes in a past master's solo than a lower-order sequence (such as a specific sequence of two notes). Higher-order sequences, then, are more suitable for producing the appearance of stylistic fidelity, yet they run the risk of resulting in texts that seem to be more fixed, predictable, and, indeed, exact copies of a master's specific solos rather than as new solos in this master's style.[6] Lower-order sequences, on the other hand, are better suited to producing variety, but they run the risk of jeopardizing stylistic recognizability. Randy commented on this dilemma and how team members tried to solve it:

> There are times when Syrus would be more strict in sticking to the chains [sequences], whereas when you lower the order you get more variation, which can be more interesting, but then you are not necessarily as close to a true representation of the style. The likelihood that you are hitting an event that's something the master would have done increases in high orders. So we did a variable order that kind of gives the best of both: at times it could introduce more chance and at times it could strictly adhere to the same events they [the masters] would have chosen.[7]

Sterne argues that "when sound-reproduction technologies barely worked, they needed human assistance to stitch together the apparent gaps in their ability to make recognizable sounds" (2002:246), and that consequently "people [delegated] their skills to technology in order to help it work" (247). James's research team provided Syrus with the same kind of assistance. Syrus's ability to mediate and reproduce styles (as a condition of possibility for later mixing those styles) was supported by team members who made sure that those styles would be suitable for reproduction and mediation. In the process, team members effectively created the styles that were supposed to exist independently of, and prior to, their mediation or reproduction by Syrus.

The steps taken by the research team in the process of designing Syrus point to one of this process's unintended consequences. James was motivated to design Syrus because he wanted to inspire musicians by creating new and hitherto unprecedented improvisation styles. The functionality of the technology he designed to achieve this goal depended on the radical simplification and hence misrepresentation of the improvisation styles of the past jazz masters that he no longer found inspiring and consequently wanted to replace. James's pursuit of future sources of inspiration has thus undermined to some extent existing sources of inspiration and thereby contributed to the problem that had originally motivated him to design Syrus.

Primed Audience

Sterne argues that another reason sound fidelity became an uncontested notion was the complicity of listeners, who wanted to believe in the ability of sound-reproducing technologies to deliver such fidelity and consequently ignored or explained away any evidence that they might not be able to do so. The commercial companies that manufactured sound-reproduction technologies educated listeners to adopt this stance: "Sound fidelity was, ultimately, about faith and investment in these configurations of practices, people, and technologies. . . . Throughout the early history of sound media, performers and listeners lent some of their own mimetic powers to the machines so that they might be dazzled. In developing their audile technique, listeners learned to differentiate between sounds 'of' and sounds 'by' the network, casting the former as 'exterior' and the latter as 'interior' to the process of reproduction. They had to be convinced of the general equivalence of the live and the reproduced" (Sterne 2002:283).

Similarly, we should not take at face value the kind of "aha effects" (Pachet 2002:79) that are produced when musicians and audiences seem to recognize their and others' playing styles in the playing of technologies such as Syrus, David's system, or The Continuator. Such effects frequently depend on the audience's desire to believe in these technologies' ability to reproduce or mediate those styles. Indeed, James's team members explicitly acknowledged this, as well as the fact that they took this desire into account when they developed Syrus's architecture. "Look," Randy told me, "there's what the average Joe wants to hear when you say, 'OK, listen, Coltrane and Miles Davis are being combined.' You know, the rhythmic activity in Coltrane, they want to hear that. And they want to hear some of the voice leading of Miles Davis. They want to hear these key elements that we want to associate with them." Not only were team members aware of the audience's desire to believe, they actively designed well-crafted demo performances in which Syrus demonstrated what team members thought the audience wanted to hear. Randy continued: "We had to focus on these demos, and while they are also interesting from a research perspective, you have to make the obvious things even more obvious. So it became that kind of conflict between doing it for the sake of being interested in the ways in which these things [styles] can really combine, and making it for the sake of, making them combine in a way that's really obvious to the average person that the two things are being combined."[8] Rick, another team member, suggested that had the audience not known in advance that Syrus's playing was supposed to be "in the style of" Thelonious Monk or John Coltrane, they would not have associated Syrus's playing with these masters: "'Playing in the style of' means more than traversing in the Markov chains of these notes and these rhythms. There is so much more to it than that. The algorithms' naivety doesn't merit the phrase 'This is in the style of Monk.' No one would hear the music that Syrus makes in the first place and say, 'This is in the style of Monk.' Never. It's not necessarily a bad thing. It's like an advertisement. It's like, 'Hey, come and take a look at this.' It's to get people in the door, I think." Matt similarly commented: "We always present it in strict terms, like it's 30 percent this, 30 percent that. But you don't really perceive it as the Coltrane style and 'This is precisely what it is.' It's more the statistical probabilities and what they're doing. We combine data from two or three or more artists into a sort of one statistical probability. The ways we express it is just something that we got used to after demoing Syrus, giving an easy way for people to understand what is actually going on and how we are combining statistical databases of these artists."

Among the team members there was disagreement about the wisdom of claiming that Syrus can not only imitate but also mix the styles of well-known jazz masters. As Rick described it to me:

When you are premising Syrus's playing by saying that it's 50 percent Coltrane and 50 percent Monk, you sort of already build this image in the listener's head. And that might not be the best thing to do. Because, you know, when I first heard that I was like—"Oh my god, I'm going to listen to this fabulous thing, the two greats have come together"—and I was really underwhelmed with what came out. But probably with no introduction at all, if I had heard it I would have approached it very differently. Like last year during a university event, Syrus performed with Matt who was playing the drums, and everyone absolutely loved it because there was no introduction about Thelonious Monk or John Coltrane and it was just a robot playing along with a human drummer. It was very nice and it was making good music as well and I think that even though it was precomposed, my favorite pieces with Syrus are those that are partly precomposed. I think that the real thing that's missing in Syrus right now is the improvisation, that sort of hook or personality.

Rick's comments are noteworthy for three reasons. First, Rick acknowledges that Syrus's "architecture" also includes the strategic priming of the audience to believe that Syrus has the capacity to play in a style that is a mix of the great jazz masters' styles. Such priming encourages the audience attending Syrus's performances to "assist" it to accomplish what it is supposed to be able to do independently. Second, however, Rick argues that precisely because this priming sets such a high standard, it might backfire, because of the discrepancy between Syrus's actual capabilities and this high standard. The result can be a severe disappointment, as Rick himself experienced. Finally, note that Rick argues that Syrus works best as an instrument that plays precomposed pieces—a claim that brings us back full circle to the reality that James wanted to escape, that is, the robot as an imitative instrument rather than a creatively generative musician.

A number of team members acknowledged that they were underwhelmed by Syrus's playing. For example, Matt commented that "I wasn't amazed by it. I don't know anyone who would have been fooled." Only Randy granted Syrus a modicum of success: "I haven't heard any of the masters in Syrus's playing. Once in a while he would go to one of those phrases where you could say, 'Oh, I can see the connection.' But that's it." It is significant, then, that although none of the team members actually thought that Syrus had the capacity to imitate any of the masters' styles, they were excited about Syrus's playing when it was supposed to be a mix of these styles. Recall that James told me that "combining and morphing different styles" has not been done before and that "this is our novelty. In this way we can generate responses that Chick Corea would have never thought of because suddenly it's 60 percent Chick Corea and 20 percent Miles Davis and 20 percent you." The promise of mixing and combining the styles of different jazz masters and thus creating a "supermaster" or "superstyle," as it were, was so alluring that team members had no difficulty embracing the belief that Syrus might be able to inspire its listeners by playing in a style that is a mix of different styles, although they did not believe it was able to convincingly simulate any one of those styles in an unmixed form.

When I asked Randy how successful the combination of styles was, he provided a clue to what made this contradiction possible: "I think it was great. That was probably the topic that interested me the most about that research. Not necessarily as to whether it was accurate in morphing the two styles of Monk and Coltrane, because obviously it's tough to say because we don't have a point of comparison of what a morph of the two should sound like. But it definitely created some interesting results that were a lot of fun to play with. And it felt new. It wasn't something you heard before. So that was the part that I found the most interesting in the research." His comments suggest that because a listener cannot compare Syrus's playing to anything he or she might be familiar with when this playing is supposed to represent a mix of styles, he or she is less likely to be disappointed and more likely to be excited while listening to this playing. They validate Rick's prior observation that priming the audience with the expectation that Syrus can play in the individual styles of specific jazz masters could easily backfire because the audience would quickly realize, based on their familiarity with those jazz masters' individual styles, that this was just what Syrus could not do. James told me something similar over lunch: "The way I try not to fail is by not trying to model a single player and the human ingenuity but rather to push it to other places, like mixing styles or like trying to show how Syrus can play like a robot." Pitching Syrus as being able to play in a style that is a mix of the masters' styles without first proving that it can convincingly mediate or reproduce any of the masters' styles in isolation thus allows the research team to prime the audience (and themselves) to believe that Syrus can do unprecedented things, because such a pitch deprives the audience (and themselves) of any standard against which they might be able to evaluate the veracity of this belief.[9]

"Thinking" like a Machine: Computational Complexity's Complications

In addition to creating new inspiring styles by mixing the improvisation styles of different jazz masters, James hoped to create inspiring new styles by adding different kinds of generative computerized algorithms to Syrus's architecture. In a brainstorming session with his research team, James said:

> For the improvising, we want to do something that humans will not do. In this case, it's to combine the styles of different jazz masters in a very algorithmic way ... which humans will not do, at least not in this way. But then what we also did is genetic algorithms, which is also something that humans don't do—they don't calculate fifty generations of mutations and crossovers in less than a millisecond. We also tried fractals a little bit but we didn't push it enough. Why fractals? We are looking for something that has some promise in creating patterns that are interesting.... But I need you to think: what other interesting things can you think of that will inspire us?

In addition to Markov models and genetic algorithms, James mentions fractal-based algorithms, which have been used for decades to generate music that is self-similar in structure across different scales or levels of magnitude (Nierhaus 2010). All of these different kinds of algorithms generate music based on complicated calculations that incorporate both stochastic and deterministic (nonlinear) processes into their logic. James attributes their added value to the fact that they can generate musical ideas that humans never could because of their more limited computational power. His comments reveal his belief that inspiration and creativity can be mapped onto algorithmic calculation, and that one way to increase inspiration and creativity lies in more advanced forms of algorithmic calculation that surpass humans' cognitive abilities.

However, this belief in the power of computational complexity to inspire humans, which is based on a very limited idea of what they find inspiring in jazz music, could also backfire. Although Syrus produced music that was always contextually meaningful because it was always generated in relation to specific tunes and their harmonic progressions and in response to the musical ideas of the musicians who played with it, the team's emphasis on computational complexity often resulted in a musical output that was far removed from the many dimensions that provide the basis for meaning in jazz, which cannot be neatly mapped onto complex algorithmic calculations. Consequently, listeners often had a hard time relating to Syrus's playing.

For example, at the beginning of my fieldwork, Matt demonstrated Syrus's different features to a number of students and visitors to the lab. At one point he put Syrus in a turn-taking mode and then played a very short phrase on the electric keyboard connected to the computer that animated Syrus. After a few seconds, in response to this phrase, Syrus's four arms moved rapidly across the marimba and played a very long and convoluted phrase. After it finished playing, we all stood in silence. Rick, amazed, asked Matt, "That was in response to what you did?" Everybody laughed. Matt said, "Yeah, sometimes it's really out there." Rick continued: "The reason I ask is that, to me, I can't see the relation to what you played." In a later conversation, Rick said to me:

> I think that because Syrus plays like a machine, the human brain is sort of not wired to completely understand what it's doing because of the algorithms that it uses. Sometimes it plays phrases that a human would never play. Which you can consider inspiring, but, to me, . . . I hear a lot of haphazard phrases that don't seem to have any sort of intrinsic melody or pattern and I don't think it's a shortcoming of Syrus. I think it's doing something so advanced due to the algorithms, that it's completely disregarding the limitations that humans play with, and because of that I think that as a listener I am able to connect less with Syrus because, you know, years of listening to music which is based completely on human limitations sort of accustom your mind to expect these things, and your human expectations, what you would describe as pleasant music is completely based on human limitations. . . . So when someone completely surpasses all of that—to me it's analogous to those players who—you have probably like a handful of artists who are so advanced that only the highest level of musicians can understand what they're doing and appreciate them. Anyway, most of the time I cannot understand what Syrus is playing. And I hear similar comments from other people here—"It is playing something I cannot understand."

Rick argues that Syrus produces music most of its human listeners cannot relate to. On the one hand, his comments expose the limitations of the team's belief that inspiring new music can be generated by means of more computationally advanced calculations, and that inspiration and creativity can be mapped onto and generated by such algorithmic calculations to begin with. On the other, his comments reproduce this belief, inasmuch as he conceptualizes the nature of the problem in terms of humans' cognitive limitations vis-à-vis Syrus's advanced calculations.

That listeners frequently experienced Syrus's playing as meaningless is noteworthy for other reasons as well. First, recall that James designed Syrus to escape the uninspiring contemporary musical landscape that he associated with the academization of jazz training. James argued that many academic programs train hundreds of musicians to play in the same ossified styles and by means of the same standardized methods and pedagogy that fetishize abstract theory at the expense of jazz's real-time emergent properties. However, because Syrus's playing is frequently determined by complex calculations that produce musical results listeners find difficult to comprehend, Syrus's end results are eerily similar to what some critics of academic jazz education, such as the saxophonist Branford Marsalis, call "think-tank music" (Young 2006)—music so brainy that it leaves its listeners either baffled and frustrated or indifferent. Rick's argument that Syrus is "analogous to those players who—you have probably like a handful of artists who are so advanced that only the highest level of musicians can understand what they're doing and appreciate them" is virtually identical to Marsalis's criticism of the music that musicians who are trained in academic jazz programs produce: "Today's jazz musicians are too mathematical and wonkish. . . . Jazz clubs are half empty, only frequented by other musicians who appreciate each other's showmanship. Listeners need music degrees to understand what they're playing" (Olding 2019).

Second, recall the specific kind of inspiration that James was looking to create by building a fully embodied robotic musician. James criticized people like David who use computerized improvising systems as mere instruments generating the very kind of music that has been played for years: "My purpose is not to build a tool for you as a musician, but rather a computer that is a musician by itself. It's the difference between the computer as only an instrument for you to use and play with and the computer as a musician that activates you and that can really inspire you." However, when musicians traded fours with Syrus or engaged in other kinds of musical interaction with it, the end result was frequently a lack of interaction, because the musicians had trouble relating to Syrus's playing. In a conversation I had with Randy, I asked him if a human musician could find inspiration in Syrus's playing. He responded:

> Yes, but it takes a certain kind of player. For instance, someone who is interested in free jazz would undoubtedly be interested in what Syrus does. [After I asked him to explain why a musician interested in free jazz would be Syrus's ideal playing partner, Randy said:] I think that some people who are trying to play with Syrus—for instance, I am playing the keyboard, and I and Syrus are trading fours, and then I am playing something that is from a standard and I'm constantly swinging and then I stop, my phrase is over, and then Syrus plays and you can sort of tell that it's in the same harmonic framework that I'm in but what it is playing is something different, obviously. So here I am, see Syrus, Syrus stops playing, and then I'm swinging again on the standard and that's—I mean, that's not good. That isn't what it's for. That hypothetical person is just playing for the novelty of playing with a machine. He isn't actually listening to what Syrus is doing, which is totally out there [musically]. And I think that people stepping up got to be more open to this sort of thing. I don't ever hear somebody playing with Syrus, trading fours with it, and then adapting their style to be more jagged and less scalar like Syrus's style of playing. They don't adapt to be like Syrus. They're expecting Syrus to adapt to be like them.

Randy's eyes fell on a poster that was posted on one of the lab's walls, which announced an upcoming joint performance of a well-known hip-hop musician and Syrus. He pointed to the poster and said: "In a few days we'll have this artist coming up, a great hip-hop artist. And that's all good, but why aren't we inviting a free jazzer? Why aren't we inviting a person who doesn't like to groove in this human sense? We need someone who is more experimental. Honestly, I honestly want to see someone free play with Syrus and start playing like it, or start playing more like it."

Randy argues that because Syrus's playing is so "far out"—that is, it appears to have a tenuous connection to the tune—a person who is a straight-ahead player will have a harder time engaging in a meaningful musical interaction with it than someone who is used to playing free jazz, a style of improvisation in which players' main point of reference is one another's playing rather than a given harmonic progression, and in which they are less tethered to mainstream jazz's stylistic conventions. It is noteworthy that he describes the kind of musician that the research team should invite to play with Syrus as "a person that doesn't like to groove in this human sense," that is, a person who can play more like the machine that Syrus is. Randy's description of the frustrating lack of interaction between Syrus and its human interlocutors is uncannily similar to David's description of his motivation for building his system. Recall what David told me about his system—that it "will always be playing inside the changes, and at heart I am an inside player. Many times in a jam session, if you are out there and the tenor player is playing with you on this tune, he's honking and squeaking and playing far-out stuff, and you're trying to sound like the early Miles, you know. It doesn't work very well. There's no conversation. It's two guys talking about different things and past each other." David is describing a situation in which a straight-ahead player (i.e., David, who is trying to play like the early Miles Davis) is trying to play with a tenor player who is "playing far-out stuff" akin to Syrus, with the result that the two players simply talk "past each other." Whereas David has managed to escape from this frustrating situation by building the kind of system that James criticizes, James's system has ended up reproducing this situation.

Third, some members of James's research team argued that in order to remedy this situation, Syrus's playing needs to be simplified and brought down to the level of complexity displayed by systems such as David's. Rick explained:

> What you have now is the equivalent of an advanced jazz musician playing with someone who had just started learning how to play the piano. Essentially it's that kind of mental difference. You can't play something interesting with him—it's like a child who is playing "Happy Birthday to You" and you are like, "What?" So I think that if you bring down Syrus's level and say, "You know what, I'll take all that information but I'm really going to limit how I analyze it and what I do with it"—not limit it in a bad way but limit it in a more human way—that might sort of make things more interesting.

Whereas Randy, whose commentary I discussed earlier, argues that the research team needs to find those few players who can adapt themselves to Syrus's playing by playing in a less "human" way, such as free jazzers, Rick argues that the research team needs to limit Syrus so it can play in "a more human way" and thus become relevant to a broader group of human musicians. Although Rick and Randy offer different solutions to the same problem, both of their solutions are based on the same ideological notion that inspiration and creativity are purely algorithmic and cognitive processes. They conceptualize both the problem and the solution to it in purely mentalist and computational terms. This dynamic exemplifies how experts succeed in absorbing critiques of their knowledge—even critiques that they themselves raise—"back into the realm of expertise" (Li 2007:11)—in other words, how the failure of expert knowledge to deliver on its promises can still support it if this failure is conceptualized in the same framework that this expert knowledge provides.

Sounding and Looking like a Machine: Embodiment's Potentials and Pitfalls

To build a jazz-improvising system that would encourage its human interlocutors to engage with, explore, and be inspired by its new musical ideas, James decided to design Syrus as an acoustic instrument-playing robotic musician rather than as a digital music-producing software that uses a tone generator such as David's system. Syrus was designed to have a long "neck" and a "head" at the end of the neck, which could bob in synchrony with the music it played. At the center of the head was a dark surface that functioned as an "eye" that could "blink" and that hid a motion-sensitive camera that could track and allow Syrus to follow the movements of the human musicians who played with it. Last, Syrus had four "arms" that could hover horizontally above the marimba bars, each arm equipped with two mallets. The rest of its machinery remained hidden from view by the marimba, so that all the musicians playing with it, as well as the audience facing it, could see were its neck, head, eye, and arms. The research team hoped that Syrus's embodied form and the different embodied cues that it could send (and perceive) by means of it would result in a tighter interaction between it and the musicians playing with it, who would thereby be more likely to be inspired or "activated," as James put it, by Syrus's playing.

For example, when Syrus improvised it would lower its head as if it were looking at its arms and the marimba bars, whereas when the human player improvised it would move its head to look at him or her. These modes created a matrix of visible interactional possibilities between Syrus and its human interlocutor, as shown in figures 3.2–3.5, which are still pictures taken from a video recording of a joint improvisation session between Syrus and me in the lab. The four interactional moments depicted in these figures took place successively in the short window of a few minutes.

According to Matt, the research team's goal was "to take the next step in computer music and apply it to a physical being like the robot so that you can visually interact with it and you're not just listening, you're not just responding to different noises [coming out of a computer], you're actually interacting. And I think that just when you embody something like that, the level of inspiration hypothetically should increase because now more of your senses are involved and you can interact with it on many different levels than you can by just listening to generated notes or something like that."

FIGURE 3.2. The author and Syrus at the beginning of their joint improvisation.

FIGURE 3.3. The author improvising while Syrus accompanies him.

Team members' experience of playing with Syrus validated these ideas. Matt, who had the most extensive experience of playing with Syrus, including in a number of public performances, commented:

> It's very helpful when Syrus would turn and look at me because then I know, "OK, this is when it's 'listening.'" And at other times, like I'll give it some of the jazz stuff and it would be helpful just to get the idea of what it's doing. So we'll run like a specific head movement function into one of the listening or note generating functions that we have. . . . Like when it was playing the melody, it would look at its arms but its head would be a little bit higher, and then when it started playing faster and more improvisation its head would go down like closer to the notes, like focusing, and it sort of gives the illusion that it is

FIGURE 3.4. Syrus improvising while the author accompanies it.

FIGURE 3.5. The author and Syrus improvising at the same time toward the end of their joint improvisation.

concentrating and knows what it's doing. So you sort of know where you are in the code and what's going on just by the head movement.

I asked Matt to give concrete examples for these helpful embodied dimensions based on his live joint performances with Syrus. He said:

> It definitely helps with synchronization. I mean, I would be able to hear if it started playing the melody when I was playing something that doesn't fit with it even without the head movements, but it helps to be able to see that, especially in the sort of circumstances when the drums are louder than the marimba and you're not in the best acoustic environment. And this happens a lot in real live performances that I've done. Like you can't really hear the piano or the bass player and everything is sort of based on head nods and eye contact. So this sort of alludes to that. So when I perform with Syrus, that's what I get the most out of because even though I know that it isn't really looking, I know the head movements so I get the idea of what processes are going underneath and what it's doing and that's the most helpful part for me.

Matt's comments are significant against the backdrop of David's description of his experience of playing with his system. Recall what David told me about playing with his system: "You need to remember the [preset interaction] setting. If I'm in a performance and I walk around people in the audience, playing the trumpet with the wireless mic attached to it . . . and I don't remember the setting and can't look at the computer screen because I am not near it, I say to myself—'Well, I'll play a little bit and see what happens.' So I'll start playing the head in case that comes up. And if it doesn't, so—'OK, I'll continue improvising.' You know, you roll with the punches." As opposed to David's system, Syrus has the capacity to send a few meaningful embodied contextualization cues that can signal to its human interlocutors whether it is playing the melody, improvising, or "listening."

Syrus's embodied features, especially its head, help engage the audience, too, not just the musicians who play with it. According to Rick, the head

> was probably *the* most successful aspect of Syrus. The moment it turns its head to listen to the human player, the whole crowd goes crazy. Even though it's not really listening with its head, just the implication that it does sort of adds all this new element because now it sort of grabs you visually as well. . . . The social aspect is completely fascinating even though right now it's very simplistic. It's just turning and nodding to the tempo detection. There is a lot of potential. I think the head is so fascinating to people because it makes the robot accessible to someone who is not technical. When Syrus responds to a musician and looks at him, it gets people through the performance and sort of leads them into the music as well because I think it affects their perception

of the music. It's a little bit like cheating but, at the end of the day, whatever works, works.

Syrus's ability to keep the audience (and, as I myself experienced firsthand, the musicians who improvise with it) engaged marks it as a kind of technology of enchantment (Gell 1994), in complete contrast to David's system. Recall that David was forced to play a lot during his live performances with his system "because given the visual thing, ... when it [the system] solos ... all that would be left is a little computer on the table . . . . The audience wouldn't know what to do with that." The fact that David's system lacks an engaging embodied form means that whenever his system improvises, the audience sees nothing but a small laptop computer and hears nothing but electronically synthesized sounds. As a result, David had to work hard to combat audience indifference—something that also made it difficult for him to secure gigs.

However, that Syrus could so successfully engage the audience by means of its embodied social cues and that it played an acoustic instrument also frequently resulted in audience indifference, but of a different kind. "Most people are very intrigued by Syrus and the head movement," Matt told me, "and the social interaction is really what they are drawn to and: 'Wow, this person is really interacting with this robot,' and they get the sense that it's listening and it is responding and the human is really interacting socially and musically with this robot. And I think many times it's more just based on what the head movement is doing and I don't really know if they're actually listening to the playing." Syrus's embodied form may focus the audience's attention on the visible dimensions of the interaction between Syrus and its human interlocutors, but it can also result in audience indifference to the music that Syrus and its human interlocutors produce in the context of this interaction.

To be sure, this indifference is also a function of the fact that Syrus's music is difficult to relate to, for the reasons I discussed earlier. However, Syrus's embodied form also plays a role in making its music difficult to understand. To illustrate this point, consider the following vignette. One day as I was checking my email in my workstation, Matt entered the lab. He greeted me and the other students who were present and then sat in front of the electric piano. I saw that Syrus was turned off, so I assumed that Matt was planning to just fool around on the piano, something many students did from time to time as a diversion from work. Matt started to play what seemed to be rather conventional musical phrases, but in a highly unconventional way. He played one note with his left hand and the next with his right and so forth in a rapid fashion, each note on a different section of the keyboard. The music he produced consequently had a zigzag contour. Suddenly, Kim's head, which up until then had been hidden behind the computer monitor in his workstation, appeared: "Syrus music!" he exclaimed with a smile. I looked at Matt and saw that he was smiling to himself while continuing to play these highly convoluted phrases.

In a conversation a few hours later, I asked Matt what Kim had meant by "Syrus music." Matt laughed and explained:

> You see, Syrus has its own style because of the arm movements and the physical limitations. You'd hear the beginning of a natural run [a phrase that consists of notes adjacent to one another] and then suddenly a note would go up in the octave—you'd hear some note being played by a different arm in a different octave because the first arm is not fast enough to play it so the other arm would compensate for it. And I think that's unique to Syrus. So Kim would sometimes refer to this as "Syrus music" and he'd play this way and he'd say, "This is Syrus music!" So today I just decided to do the same.

Matt's comments highlight the fact that Syrus's embodied form results in a specific style of playing. Although this style was frequently a source of amusement, many team members considered it to be a problem. Rick commented on it:

> Syrus tends to play in a sparse, wandering sort of pointed style. In a free style that isn't rhythmically or quite harmonically grounded as straight-ahead jazz, this wouldn't be a problem, obviously. But as most music is a product of the idea that things must be coordinated or that things must fit the rhythmic grid, it does become a problem. Syrus tends to play things that aren't scalar, which is a huge difference from people. There's something to be said for the fact that when you grow up and when we are conditioned to play music, scales are the building blocks, and that is something I really think no human can ever shake. Just from the time they were five years old it has been ingrained in them. And that is obviously something that Syrus doesn't have, and it takes a certain open-mindedness to look at what it does and take something from it.

Syrus's own style of playing serves as a reminder that fantasies about limitless stylistic flexibility and malleability enabled by digital technologies and interfaces with sliders that govern different proportions of styles will always be limited or curbed by the material infrastructure supporting such practices and technologies. There will always be some material stratum that imposes its own constraints and limitations, and these will eventually result in a specific style of styling styles in a particular domain. The constraints that result from Syrus's particular embodied features—the speed at which one arm can play successive notes, for example—produce a residual style that pervades its playing regardless of which jazz style or mix of styles it happens to be playing in at any given moment. Everything that Syrus plays will always be in that edgy and difficult-to-relate-to style with which team members associate it.

As I explore in the next chapter, at the same time that these and similar dimensions of Syrus's embodied form jeopardized the kind of aesthetic inspiration that James had hoped to generate by designing Syrus, other dimensions of this form functioned as unexpected sources of aesthetic pleasure. However, not only was this aesthetic pleasure different from the one that James had hoped to produce, it was the result of malfunctions in Syrus's hardware and software, rather than the product of their seamless operation.

4

Separating Noise from Signal:
The Ethnomethodological Uncanny as Aesthetic Pleasure in Human-Machine Interaction

"A Parkinson Moment"

I was sitting at my workstation in the lab, checking my email. Syrus was bobbing its head in a repetitive wide motion in the center of the room. Although Syrus was not in play mode, Matt, who was sitting at his workstation, had not turned off the feature that was responsible for the bobbing of Syrus's head, which was supposed to be synchronized with the rhythm of Syrus's playing. Team members often left this feature on even when Syrus was not playing so that they would not need to initialize the head each time they wanted Syrus to play. During my fieldwork in the lab, Syrus's head often bobbed silently in its default rhythm as team members worked at their workstations.

Suddenly, on the periphery of my vision I noticed that Syrus had stopped bobbing its head and was instead shaking it rapidly and emitting loud, rapid clicking sounds. Matt, Kim, and I looked over our computer screens. "What's going on?" I asked. "Oh, it's one of Syrus's Parkinson moments," Matt said, laughing. "It happens when the head receives conflicting messages. It's a bug we haven't figured out yet." We all approached Syrus and watched its head-shaking and listened to the clicking sounds. At a certain point, Kim started shaking his head in imitation of Syrus, and we all burst out laughing.

Some scholars might approach this vignette as an example of people's willingness to engage with sociable robots and other kinds of relational artifacts as interactional and even sentient partners. For example, Lucy Suchman argues that machine designers take advantage of specific features of human conversation to encourage humans to attribute interactional capacity to the machines they use. Such features include humans' ability to engage in "the detection and repair of mis- (or different) understandings" (Suchman 2007:12). A classic example is the ELIZA programs that Joseph Weizenbaum designed in the 1960s to explore the possibility of natural-language conversation with a computer. Weizenbaum programmed the computer to respond to the queries of the user *qua* patient as if it were a Rogerian psychotherapist. Even when the computer generated responses that appeared nonsensical, users reasoned that they must be motivated by some psychiatric intent (Suchman 2007:48– 49). Similarly, based on her research on robotic companions, Sherry Turkle has argued that "as early as the ELIZA program, both adults and children are drawn to do whatever it takes to sustain a view of these robots as sentient and even caring. This complicity enlivens the robots, even as people in their presence are enlivened, sensing themselves in a relationship" (Turkle 2011:85). Turkle's informants, children and adults alike, normalized and repaired the misunderstandings and gaps that resulted from malfunctions in the sociable robots they interacted with. When Kismet, a sociable robot developed at the Massachusetts Institute of Technology, suddenly stopped functioning, children reasoned that it was sleeping. When it produced random sounds, a child interpreted them as a message directed to him personally (91). Kids argued about a robot that failed to respond to their queries (because of a problem with its microphone) that it was not fluent in English (87). These reactions evince people's eagerness to repair robots' rudimentary performance of sentience and to bridge "the gap" between such a performance and their own desire to have a digital companion (25; see also Katsuno 2011; Richardson 2016:116, 124).

Matt's choice to call Syrus's condition "a Parkinson moment" is a borderline example of this stance. Although it first appears to "fill in the blanks" (Turkle 2011:24) in Syrus's behavior and thus to extend somewhat the interaction with it, it has a different meaning. "A Parkinson moment" is a moment in which the body, which in the context of a predominantly Western—that is, culturally specific—ethnopsychology or local theory of mind is understood as the primary medium of human existence, something that is supposed to remain in the background as a kind of invisible channel in the service of an intentional and immaterial self, comes to the fore and becomes the focus of attention because it is malfunctioning. Such a moment reveals the body's materiality, undermining the culturally specific notion of an intentional self in control of its actions. This was the case with Syrus. If my interlocutors in the lab were willing to relate to Syrus as a living being, it was mostly through instances in which human life appears in its most medium-based form, lacking in intentionality because the human body was failing to function properly.

Furthermore, rather than trying to repair the deficiencies in Syrus's behavior by ignoring or rationalizing them as the behavior of an intentional entity, team members focused on those gaps, explored their material basis, and took pleasure in widening them. They maximized the gaps by animating Syrus in ways that only on the surface resembled a stance committed to repairing the interaction with it. In other words, far from detecting and repairing Syrus's communicative inconsistencies or working around its communicative limitations, team members were more intent on producing further damage.

Suchman, who pioneered the anthropological and sociological study of human-machine interaction, was heavily indebted to Harold Garfinkel's ethnomethodological analysis of social interaction. A key source of inspiration for Garfinkel's theory, one that up to now has gone more or less unacknowledged, was cybernetics. Cybernetics was heavily invested in theorizing how homeostasis and order can be maintained through negative feedback mechanisms, which can compensate for noise and disorder in self-regulating systems. Garfinkel's theory of interaction consequently emphasized as a normative ideal what I call "interactional homeostasis," that is, the idea that participants in an interaction strive to maintain interactional order and compensate for interactional noise and disorder through negative feedback mechanisms such as "repair work." This intellectual legacy is why a key strand in the aforementioned research on human-machine interaction has emphasized humans' investment in detecting and repairing machines' communicative inconsistencies and gaps in the course of interacting with them.

But Garfinkel's theory, and the theories of human-machine interaction that were informed by it, run into trouble when confronted by the Romantic, counter-Enlightenment aesthetic category of the uncanny, which, as I argue in this chapter, turns on interactional noise and disorder as normative ideals and desirable goals. Whereas the uncanny has often been attributed to uncertainty about the animate or inanimate status of an entity, such uncertainty is an epiphenomenon of a more fundamental process, namely the sudden awareness of the materiality of the semiotic forms most intimately linked to the self, such as one's house, body, voice, and language, which under normal conditions are supposed to remain invisible in the context of culturally specific theories of mind in the modern West. As in the "Parkinson moment," this kind of awareness results from the occasional noisy materiality of semiotic forms, which makes them reflexively point to themselves. Such noise can become a focus of attention, an object of fascination, and a source of aesthetic pleasure against the backdrop of, and in reaction to, powerful Enlightenment-era semiotic ideologies that have attempted to suppress the materiality of semiotic forms. The uncanny restricts the applicability of Garfinkel's theory of social order. And since the uncanny has often been observed in the context of human-machine interaction, it undermines the idea that humans are inclined to repair machines' communicative inconsistencies and gaps that emerge while interacting with them.

To understand how the kind of stance displayed by my interlocutors with respect to Syrus has remained unacknowledged and undertheorized in the social-scientific study of human-machine interaction, it is necessary to unpack the intellectual history of some of the scholarly views that have come to dominate this field. Once unearthed, this intellectual history can open up the conceptual space needed in order to develop an alternative theoretical framework, in the context of which many of the instances of human-machine interaction I observed in James's lab—such as the one with which I opened this chapter—can start making sense. The first half of this chapter focuses on unpacking this intellectual history and on theory-building that turns on the notions of the uncanny and animation; the second half presents an analysis of my interlocutors' interactions with Syrus in light of these theoretical insights.

The Cybernetic Foundations of Ethnomethodology

One of the key goals of Garfinkel's ethnomethodological analysis of social interaction was to understand how people make sense of "what a person is 'talking about' given that he does not say exactly what he means" (Garfinkel 1967:78; see also Suchman 2007:80–84). Garfinkel argues that mutual intelligibility in interaction is the product of strategies or ethnomethods of sense making that interactants enact during interaction. Those methods include, for example, turn-taking conventions that produce an orderly background on the basis of which interactants can make sense of and mutually orient to the interaction's contingently emergent referents. Participants come to the interaction committed to using such methods of sense making and to producing and maintaining interactional order; they also assume that their interlocutors are intentional entities. They construct themselves as trusted and intentional interactants vis-à-vis other interactants by competently enacting these and other methods.

Scholars in conversation analysis—a strand of research that emerged directly from ethnomethodology—have elaborated on this commitment by empirically analyzing "repairs," that is, how participants negotiate trouble that emerges during the interaction and that can undermine the interactional order. In the modern Western conversational contexts that were the focus of the foundational studies in conversation analysis, interactants tend to self-correct. When other-correction takes place, it is frequently modulated in form (Schegloff, Jefferson, and Sacks 1977:377–381). Other-correction can find expression in, for example, ignoring one's interlocutor's nonsensical contribution, silently rationalizing it, tactfully providing the other person the opportunity to repair it in a subsequent turn, or, much less frequently, directly and explicitly repairing it. Individuals use these methods because they are socialized to do so.

Based on Garfinkel's work, one of Suchman's key insights was that designers of certain machines capitalize on human interactants' commitment to real-time interactional sense making and on their tendency to assume that an observable behavior must be the result of an underlying intent. For example, when interacting with one of the ELIZA programs, DOCTOR, "computer-generated responses that might otherwise seem odd were rationalized by users on the grounds that there must be some ... intent behind them" (Suchman 2007:48–49). In real-time interactions with different kinds of machines, human interactants maintain interactional order by performing different kinds of repair to compensate for the machines' poor capacity for interactional sense making. Indeed, human interactants are so committed to maintaining interactional order that they perform repair even for machines that provide very rudimentary input, such as photocopying machines (Suchman 2007:125–175; see also Orr 1996), although if such an input becomes too nonsensical, they cease to do so, resulting in interactional breakdown (Suchman 2007:161–167; see also Turkle 2011:87).

Originally published in 1987, Suchman's book was revolutionary because it extended Garfinkel's theory of human interaction to human-machine interaction. It also, however, inadvertently returned to an unknown intellectual source of Garfinkel's theory: cybernetics, a field that emerged largely in the context of the study of human-machine interaction (Wiener 1982 [1948]:5–7). It consequently reproduced cybernetics' emphasis on homeostasis and on negative feedback mechanisms as the means of maintaining it.

Garfinkel's early essays, written between 1949 and 1952 but not published until 2008, provide strong evidence of his debt to cybernetics.[1] They represent one of the most elaborate attempts to use cybernetic ideas as the basis of a sociological theory of social order. In a memo Garfinkel wrote in 1952 for the Organizational Behavior Project at Princeton University, he surveys "various conceptions of information," including, among others, those developed by Claude Shannon, whose *Mathematical Theory of Communication* was a groundbreaking text in this intellectual lineage; G. A. Miller, who developed a theory of language and communication based on information theory; Norbert Wiener, one of the key architects of cybernetics; and Gregory Bateson and Jurgen Ruesch, who applied cybernetic ideas to psychiatry. Garfinkel emphasizes two key principles that emerged from these theories (Garfinkel 2008:102–109). First, information is a spatially or temporally transmitted pattern whose maintenance requires effort, given the tendency of entropy to push any system toward the default state of randomness or noise in which all states are equally probable and no pattern can be detected. Second, in mathematical terms, information is the patterned occurrence of signals that has no meaning beyond the statistical distribution of such signals. Based on these and other axioms, Garfinkel attempts to develop an information-based theory of social order. He argues that for meaning to exist, individuals must be able to "idealize" signals, that is, to view them not as signals in themselves but as signs that carry meaning. People match signals to meanings (which thus become "signs") by following "rules" that "include not only the usual sense of grammar, but matters of temporal order, context of meanings, redundancy built into the statistical structure of language, expectations that are entertained without question" (Garfinkel 2008:150). Thus, rather than being a purely cognitive or logical affair, meaningful information is "constituted . . . as information by the social (cooperatively ordered) aspects of the situated social orders in which it occurs" (Rawls 2008:13).

Garfinkel's theory was heavily influenced by the cybernetic focus on systems that can self-regulate through negative feedback mechanisms, which oppose a system's deviation from a specific desired state. Norbert Wiener gave the classic example of a thermostat, which helps maintain a desired room temperature by activating the heating system if the room temperature falls below the desired level and turning it off if the temperature rises above it. In this case, the "feedback tends to oppose what the system is already doing, and is thus negative" (Wiener 1982 [1948]:96–97). Garfinkel (2008:209) uses the example of the thermostat too—as a model for the strategies that interactants use to correct interactional errors and to maintain what I call "interactional homeostasis" as part of their competent participation in self-regulating social systems. He argues that "we have . . . the notion of incongruity as any experienced disparity between an expectation and an outcome. Insofar as the incongruity has the effect of impugning the integrity of an order—that is, its occurrence impugns the criteria of organization . . . we shall refer to the experienced discrepancy as the experience of error" (180). Given the possibility of such incongruities or errors, "every net that is set up to accomplish more than trivial tasks will have a proliferated set of communicative devices whereby the routine aspects of the net are maintained as well as a proliferation of communicative devices whereby incongruity is controlled" (181). In a different paper he defines "the control functions of communicative work" as "the effects of communicative work for the regulation and resolution of incongruity" (211). Such "communicative work" consists of devices whose role is to minimize the possibility of error and confusion and to correct the interactional system when they do occur. They are, in effect, negative feedback mechanisms.

The cybernetic roots of Garfinkel's theory of social order also come to the fore in his definition of incongruity in terms of "noise," or "the randomness or meaninglessness of a set of signals—the unidealizable character of a set of signals" (Garfinkel 2008:183). Individuals always experience the world as noisy to some degree. Hence, the possibility of confusion is always imminent. The social scientist's task is "to spell out the conditions of [the individual's] make-up and the conditions of the net under which noise in a given amount makes a difference. And the problem that lies just beyond this is that of learning about the devices that can be used to reduce or magnify this amount" (184). Although in this quote Garfinkel acknowledges the possibility that interactants might choose to increase the "amount" of noise in a communicative event, in most of his writings he approaches interactional homeostasis as a default normative ideal. He argues that "part of the communicant's communicative work and 'internal action' will be devoted to minimizing the amount of confusion or meaninglessness. Or, in terms of the social system, much communicative work will be oriented to the maintenance of interpersonally valid definitions of the situation" by means of ethnomethods (ibid.). This is why he named his experiments, in which he deliberately produced confusion in order to observe and bring into relief interactants' ethnomethods of sense making and repair, "breaching experiments."

Although Garfinkel's heavy debt to cybernetics became less apparent in his later writings, the latter are indebted to the former. For example, Garfinkel orchestrated experiments in which he gave students a series of yes-or-no answers to the questions they posed to what they believed was a human counselor. The answers were in fact generated in advance on the basis of a series of random numbers (Garfinkel 1967:80). Here, too, the goal was to show that students would go to great lengths to subsume the randomly generated responses (*qua* noise) under the notion of a coherent intentionality that was responsible for giving the answers (*qua* signals). The experiment showed that human participants are heavily invested in separating signal from noise even when there is no signal.[2]

On the one hand, Garfinkel's insights provided the basis for understanding how humans compensate for the extreme interactional limitations of machines. On the other, his emphasis on interactional homeostasis as a normative ideal has occluded the fact that interactional noise, including noise emerging from the material infrastructure of machines, is often a source of fascination that humans strive to prolong rather than repair. In such cases, humans are more invested in separating noise from signal than in separating signal from noise, as it were. Such is the case of the "Parkinson moment" with which I opened this chapter, and of other similar moments that I observed in James's lab. Rather than being an outlier, this moment belongs to an important aesthetic category in the modern West that begs for anthropological clarification—and that presents problems for the theories of human-machine interaction that were informed by Garfinkel's theory.

The Uncanny and "the Reduction of Messages to their Signal Character"

For Garfinkel, people's worlds would remain organized as long as their experiences "meet the criteria of continuity, consistency, compatibility, temporal continuity, and clarity" (Garfinkel 2008:185). Noise can disrupt the conditions of possibility for a meaningful world by undermining these criteria. The result is "the reduction of messages to their [meaningless] signal character" (187), followed by confusion and anxiety. Garfinkel calls this situation "a state of anomie or randomness" (182), the opposite of "complete normative integration," in which "the probability of noise is zero" (183).

The phenomena that Garfinkel calls "zero-order information" (Garfinkel 2008:187) resemble the aesthetic category of the uncanny, which reached its apotheosis in nineteenth-century Romantic literature. The uncanny emerged from individuals' uncertainty about everything that they took for granted, especially their most intimate surroundings (Royle 2003:1). The best-known treatise on the uncanny is Sigmund Freud's (1964 [1919]) essay on it. He describes uncanny phenomena, such as the sudden and inexplicable repetition of sensory stimuli, events, and behaviors, as "the compulsion to repeat" (Freud 1964 [1919]:237). He gives the example of a person who "wander[s] about in a dark, strange room, looking for the door or the electric switch, and collide[s] for the hundredth time with the same piece of furniture" (236). This description mirrors Garfinkel's description of the disruption of the aforementioned criterion of "continuity" that finds expression in "compulsive repetition" (Garfinkel 2008:185) or in "symptoms of blocking," in which one's activity loses its flow and becomes frozen in "an unending moment" (186). Freud's description of the uncanny sensations experienced upon witnessing a lifelike automaton (Freud 1964 [1919]:226–228) mirrors Garfinkel's description of the disruption of the criterion of "consistency," that is, when "the specifications [of an object] are experienced as logically and/or factually incompatible. Behaviorally, the failing of this criterion is found in complaints of uneasiness about the irreality of the witnessed object . . . [and] complaints of strangeness with and distance from the familiar, routinized, and recognizable experiences" (Garfinkel 2008:186).

The resonance between Garfinkel's notion of "the reduction of messages to their signal character" and the aesthetic category of the uncanny comes to the fore most clearly in Garfinkel's discussion of noise that might emanate from individuals' bodies in the course of interaction. This discussion is directly relevant to Syrus's "Parkinson moment":

> You address a question to a friend who has a nervous tic and is given to shaking his head, now up and down, now side-ways, but in a fashion that varies independently of what you say to him. If you ask him a question requiring a yes or no answer and he has lost his voice so that he needs to nod in answer, you may have a certain amount of trouble sorting out the informing character of the messages from their "noise" character.... In a case where no separation can be made in a signal field between the noise character of a message and the informing character of the message, we talk of the random character of the meanings that can be drawn, or a state of zero-order information. (Garfinkel 2008:167)

Elsewhere Garfinkel gives the example of "a referee in Kriegsspiel [a war game] who indicates 'no' by shaking his head [and thus] may disrupt the continuity of the player's experiences if the referee shakes his head to indicate 'no' but also shakes his head in the same way due to a nervous disorder" (Garfinkel 2008:142).

There are notable similarities between these examples and the theory of the uncanny offered by Ernst Jentsch in 1906, which Freud discusses at length in his essay in an attempt to supplant it with his own theory. Jentsch argues that uncanny sensations arise in situations that make the individual doubt whether a living being is indeed living or whether an inanimate entity is indeed inanimate (Jentsch 1997 [1906]:11). In relation to instances in which human beings produce uncanny sensations, Jentsch notes that we "can always assume from our fellow men's experiences of ordinary life ... the relative psychical harmony in which their mental functions generally stand in relation to each other," despite minor deviations (14). Yet,

> if this relative psychical harmony happens markedly to be disturbed in the spectator ... then the dark knowledge dawns on the unschooled observer that mechanical processes are taking place in that which he was previously used to regarding as a unified psyche. It is not unjustly that epilepsy is therefore spoken of ... as an illness deriving not from the human world but from foreign and enigmatic spheres, for the epileptic attack of spasms reveals the human body to the viewer—the body that under normal conditions is so meaningful, expedient and unitary, functioning according to the directions of his consciousness—as an immensely complicated and delicate mechanism. (Jentsch 1997:14)

Jentsch's description of our default belief in our interlocutors' "relative psychical harmony" resembles Garfinkel's description of the normative and default ideal of interactional order. Jentsch's description of epilepsy as a form of noise that results from a malfunction in the human body, and that disrupts this harmony and our belief in it, mirrors Garfinkel's description of the noise that might disrupt the interactional order as a result of bodily failure. Indeed, when Freud, in an essay on Fyodor Dostoevsky, discusses epilepsy, he defines it as "the uncanny disease with its *incalculable*, apparently unprovoked convulsive attacks" (Freud 1961 [1928]:179, emphasis added), thus pointing to epilepsy's "noisy" character.

The importance of the resonance between the aesthetic category of the uncanny and Garfinkel's description of zero-order information is that whereas Garfinkel assumes people will try to repair any form of trouble that emerges during interaction, a number of popular artistic genres have focused on exploring rather than suppressing uncanny phenomena as "unidealizable signals" or noise. Noise has thus been the foundation of a world and an object of aesthetic elaboration and pleasure, including in James's lab, as I will elaborate later in this chapter.[3]

The Sudden Awareness of the Materiality of Semiotic Forms

For the uncanny to exist, one must be aware of the materiality of the semiotic forms most intimately linked to the self, such as one's house, body, voice, and language. In itself, however, such an awareness is insufficient. Ideologies that suppress the existence of this materiality are also needed. It is against the backdrop of such ideologies that the awareness of the materiality of semiotic forms can be experienced as "sudden" and uncanny.

The idea of the modern Western subject has been supported by a specific semiotic ideology that emphasizes the cultivation of reason through the subordination of the materiality of semiotic forms to immaterial meanings. This ideology, most directly associated with the Enlightenment, defines the modern subject as an interiority that must be kept distinct and separate from any form of materiality, broadly defined—such as the body, ritual, received tradition, other people, and words—in order to maintain its freedom and moral autonomy (Keane 2007; Wilf 2011). Key architects of Enlightenment-era ideologies of the modern subject such as Francis Bacon and John Locke performed this work of purification. They focused their efforts on making language and experience tools of reason and scientific advancement by decontextualizing them from "specific situational, discursive, personal, social, political, and historical circumstances" (Bauman and Briggs 2003:20–21). Bacon argued that the function of speech should be limited to referencing and conveying information about the world in the most neutral and transparent way. Intertextuality, rhetoric, and especially the reflexive properties of language were to be abolished (25). Locke similarly attempted to decontextualize language as a tool of reason by "countering premodern views of words as signs of the inner nature of things, as thereby tied to mysticism, magic, and alchemy" (34). He emphasized the arbitrary nature of signs and downplayed their iconic and indexical properties (37).[4] Such semiotic ideologies have informed attempts made by Protestant missionaries to abolish local Indonesian practices that involved fetishism, ritual exchange, and ceremonial speech, that is, practices in which the self and the world are explicitly and intimately entangled with the materiality of semiotic forms (Keane 2007).

I suggest that, from an anthropological perspective, the uncanny is the return of the awareness of the materiality of semiotic forms against the backdrop of such efforts to suppress it. It turns on the awareness of the opacity rather than the transparency of mediating semiotic forms, their material and formal properties, their often recalcitrant tendency to reflexively point to themselves rather than to immaterial meanings, frequently as a result of the noise they produce. This thesis gives a more precise meaning to the argument, often made by scholars of the uncanny as a Romantic phenomenon, that it should be understood against the backdrop of the Enlightenment work of suppression, in the general sense of abolishing the occult and advancing reason (Castle 1995:8). One of the most oft-quoted aphorisms about the uncanny is Friedrich Schelling's from 1835, which Freud discusses in his essay: the uncanny is what "ought to have remained secret and hidden but has come to light" (Freud 1964 [1919]:224). The Romantic aesthetic sensibility was open to the possibility that "at any moment what seemed on the surface homely and comforting, secure and clear of superstition, might be reappropriated by something that should have remained secret but that nevertheless, through some chink in the shutters of progress, had returned" (Vidler 1996:27). For the enlightened mind the world was homely, comforting, secure and clear of superstition as long as semiotic forms remained transparent and functioned as mere conduits for the expression of previously existing immaterial meanings; Romanticism challenged this premise by displaying tolerance for and interest in how the materiality of semiotic forms can constitute rather than merely reflect meaning and the self (Wilf 2011).

Classic instances of the uncanny often turned on the sudden awareness of the materiality of the semiotic forms most intimately linked with the self. "By far the most popular topos of the nineteenth-century uncanny," which authors such as E. T. A. Hoffmann and Edgar Allan Poe explored in their stories, "was the haunted house," standing "for romanticism itself" (Vidler 1996:17). Readers found pleasure in the gradual estrangement of the familiar, safe home as it was invaded by inexplicable noisy stimuli in different sensory modalities. Such stimuli made hitherto transparent semiotic forms reflexively point to themselves to reveal their materiality. One story that Freud discusses in detail, written in 1917 by L. G. Moberly and titled "Inexplicable," begins with a sentence that immediately draws the reader's attention to the house's materiality by means of the noises it produces: "The hinges were rusty, the gate swinging . . . behind me creaked dismally, and as the latch clicked in its socket with a sharp clang I started" (Royle 2003:136). The invasion of olfactory and visual stimuli is then added to the mix. As recounted by Freud, the story is about "a young married couple who moved into a furnished house in which there is a curiously shaped table with carvings of crocodiles on it. Toward evening an intolerable and very specific smell begins to pervade the house; they stumble over something in the dark; they seem to see a vague form gliding over the stairs—in short, we are given to understand that the presence of the table causes ghostly crocodiles to haunt the place, or that the wooden monsters come to life in the dark" (Freud 1964 [1919]:243–244). This synesthetic commotion leads the reader to realize that representations of crocodiles can become flesh-and-blood crocodiles. In this kind of Romantic literature, estrangement ultimately ends up in self-estrangement; that is, the subject becomes suddenly aware of the materiality of the semiotic forms that are most intimately linked to his or her self. Thus, in one of Hoffmann's stories that Freud analyzes, titled "The Sandman," one of the story's protagonists, Nathaniel, is "seized with horror" after reading a poem aloud to himself, exclaiming, " 'Whose dreadful voice is this?' " (Royle 2003:50).

Although Freud's essay was the culmination of a long Romantic tradition that explored the uncanny as an aesthetic category, the uncanny continued to be an object of aesthetic exploration and pleasure throughout the twentieth century. For example, it was a defining feature of the modernist avant-garde, especially of the latter's emphasis on "defamiliarization" (Vidler 1996:8; see also Royle 2003:97), as well as more recently of David Lynch's movies and especially his early 1990s television series, *Twin Peaks*, in which he creates uncanny effects by means of different kinds of noise that disrupts viewers' taken-for-granted assumptions about interactional order and the natural world. For example, Lynch creates uncanny effects through distorted voices, out-of-sync bodily comportment, different forms of doubles, images that become distorted by seeming interference in the television image, and time lags in turn-taking sequences—what Garfinkel calls "communicative timing [or] . . . expectations of timing . . . and [their relation to] familiarity, strangeness . . . of the other" (Garfinkel 2008:217).

The uncanny's message is thus the medium, but not in Marshall McLuhan's sense of the medium, which changes the "scale or pace or pattern . . .

[of] human affairs" (quoted in Boyer 2007:14). Rather, this is so in the sense of the materiality of semiotic forms of which any medium is made. Such materiality can become the object of heightened awareness when those semiotic forms begin to reflexively point to themselves, often as a result of their sudden noisy materiality, and especially in cultural contexts that normatively suppress awareness of this materiality. This general anthropological definition explains why the uncanny can be experienced in relation to vastly different phenomena, from one's creaking house to a hissing vinyl record that thereby reflexively points to itself. It also explains why the uncanny has most often been associated with the human form and with humanoid machines (Mori 2012). Fundamentally, humans become conscious of their world through their bodies. The body is thus not an object in the world just like other objects. Rather, it is the medium through which other objects exist for humans, "the unperceived term in the center of the world toward which all objects turn their face" (Merleau-Ponty 2002:94). When this "unperceived term" becomes noisy in the form of a nervous tic, epilepsy, or a Parkinson moment, thereby reflexively pointing to itself, the uncanny is experienced in all its force by humans whose cultural makeup consists of semiotic ideologies that suppress the materiality of semiotic forms.[5]

Uncanny Worlds

One day James came over to listen to a few new features in Syrus's playing that Matt and Kim were working on. As we were listening, the sound of Syrus's playing suddenly stopped, although its four arms continued to hover horizontally above the marimba, as if it were still playing. Kim immediately explained that this happened whenever the velocity of the mallets attached to the end of each arm unexpectedly fell so low that they could not hit the marimba bars. The image was arresting: Syrus's four arms rapidly moved horizontally above the marimba in a movement that was rich with an unrealized potentiality that we could nevertheless imagine—indeed, could not *but* imagine. It was a moment that highlighted Syrus's materiality. This moment is a classic uncanny phenomenon. It resembles "dismembered limbs, a severed head, a hand cut off at the wrist . . . especially when . . . they prove capable of independent activity" (Freud 1964 [1919]:243). In these instances of the uncanny, the body ceases to be a transparent medium in the service of an intentional and immaterial self and instead becomes the focus of attention. Rather than immediately repairing the situation, the research team began making jokes that focused on Syrus's material infrastructure. "So soft! Microsoft!" said Kim. James then directed our attention to the sound of the motors that kept Syrus's arms moving, which was now clearly audible because the mallets did not hit the marimba bars and consequently did not mask it as they usually did. "Not Bad!" James said, laughing. "That can be a feature! Motor music!" Everyone laughed. Syrus's material infrastructure, rather than some imputed subjectivity, became the basis of our relationality with it.

Team members no doubt took this playful stance toward Syrus's malfunctions because they were its developers and therefore were especially aware of its mechanical infrastructure. But their stance was also anchored in undertheorized and overlooked dimensions of sociality, namely animation and alterity, which present problems for the overwhelming tendency in anthropology to view social life through the prism of the performance of identity. This prism "reductively aligns speakers, performances, voices, and selves" (Hastings and Manning 2004:291). It assumes that people naturally aspire to experience and convey their authentic self-identity, and that they look for signs of such authenticity in their interlocutors. The theory of animation offered by Jackie Stacey and Lucy Suchman (2012) exemplifies these assumptions. Drawing on both film studies and science and technology studies, Stacey and Suchman argue that for an imitation of life to be convincing, the human labor that supports it must be hidden. Robots' liveliness depends on masking their limits and "the otherwise invisible labours of maintenance and care that [they] require, and only humans can provide" (Stacey and Suchman 2012:28).[6] They emphasize the importance of "the disguise of technique: the making invisible of the means of representation (camerawork, editing, lighting, performance and so on)" (22), noting that "the signs through which life is (mis)recognized . . . are key" (17). Such misrecognition is similar to the dynamic of commodity fetishism in which human labor is mistaken for the commodity's independent value, nature, and existence (20–21). In this specific strand of research, then, animation depends on a kind of deception or misrecognition of the political and material infrastructure that is responsible for what appears to be alive.

When this notion of animation is presented as an overwhelming characteristic of all relationships with relational artifacts, "we lose sight of the uncanny 'illusion of life' that makes these characters appealing, of their particular blend of materiality and imagination, and of their diffuse agency" (Silvio 2010:423; see also Nozawa 2016:171–175). More specifically, we lose sight of the possibility that what people find fascinating in their interactions with "performing objects"—a heterogeneous collection that includes online avatars as well as "more tangible performance objects and nonhuman actors . . . [such as] masks, costumes, dolls, puppets, animations, automatons, machines, and robots" (Manning 2010:318)—is precisely their awareness of mediation (311), that is, of the craft, material infrastructure, and labor that support these performing objects.

To be sure, as I showed in the previous chapter, the research team I worked with was not oblivious to the many ways in which Syrus could be designed and programmed to create the illusion of authentic self-expression. Indeed, finding ways to create this illusion in order to facilitate joint improvisation between Syrus and human musicians was one of the team's primary research foci.[7] First and foremost, the team was interested in exploring the expressive potential of Syrus's head. This is why Syrus's head was programmed to bob in synchrony with the music it played at any given moment. The bobbing created the illusion of a musician who "digs the music," as Matt told me. In an introductory presentation to new team members, Matt highlighted the importance of exploring the ways in which Syrus's head could support the illusion of authentic self-expression:

> One of Syrus's main purposes is to allow us to do research on the head and having the social interaction with it. Like you can look at Syrus and it will look at you. And it is so much different than if you are just playing with the arms or with the software on the computer, so this is something you should really take advantage of. So when you are thinking about reprogramming Syrus, you should keep that in the back of your mind: how can the head react to this, how can the head react to different performers. I guess it adds a more intimate relationship between the musician and Syrus. It's not just you playing and the arms are just moving, or whatever. . . . There are several things that we found that look good and seem to be reasonable, like when Syrus is improvising it would look down and its eye would face down the marimba and sort of focus on what it's doing, and when listening back and forth in turn taking with the human musician it would look at the musician each time he is improvising. And it really does add, it adds so much to the performance, to the interaction, and it's really what you should be taking advantage of when you're programming Syrus because otherwise it's just an acoustic computer, really, if you do not include the head. So take advantage of that.

Matt emphasizes the potential of Syrus's head to create the illusion of a musician who is authentically expressing himself and intentionally participating in a joint improvisation with other musicians. He argues that musically interacting with Syrus is different from simply interacting with a computer software (such as David's system) because Syrus's responses to human players receive a humanlike embodied form that sustains the illusion of authentic self-expression.[8]

At the same time, the research team often enjoyed intervening with the software that controlled the various features of Syrus's head to produce jarring effects that immediately undermined the appearance of Syrus's authentic self-expression. They took pleasure in turning the technologies of immersion they perfected into technologies of estrangement. For example, one of the team's frequent sources of amusement was to instruct the computer to slowly rotate Syrus's head vertically 360 degrees on its horizontal axis while Syrus was playing. The sight of Syrus's head completing a full circle while its arms continued to engage in an activity that is associated with deep human expressivity such as playing music never failed to produce laughter in everyone who was present.[9]

Similarly, they would often take advantage of Syrus's limited interactional competence to produce interactional noise that they found hilarious. Kim would occasionally sit in front of the electric piano hooked to the computer that animated Syrus, put Syrus in a turn-taking mode, and begin to play a classical piece. As I discussed in the previous chapter, he frequently did so to test Syrus's capacity to respond to his playing in the different styles in which it was trained (Wilf 2013d:726). He would, however, often also do so to produce comic effects, especially by playing Western classical music. The laughter, I soon learned, was not only the result of the jarring juxtaposition of some of the pinnacles of Western classical music such as a piece by Chopin, played by Kim, and Syrus's responses on the marimba, which, while bearing some relation to the music Kim played, were utterly different from it in terms of expressivity and aesthetic quality. It was also in response to the fact that when Kim wanted to produce comic effects, he would play the classical pieces as if he were playing by himself, ignoring Syrus's responses and not adjusting to them, as he usually did when he tested Syrus. By not trying to repair the gaps in Syrus's responses to his own playing in the form of, for example, adjusting his playing to whatever Syrus was playing, Kim brutally exposed Syrus's complete interactional inflexibility and its inability to perform any kind of repair. This resulted in interactional nonsense and disorder that delighted everyone. By continuing to play, Kim exposed Syrus's turn-taking mode for what it really was: a mechanically rigid make-believe that achieved its effects only when Syrus's human interlocutors adhered to a very clearly defined turn-taking sequence and continually repaired Syrus's turn-taking lags and errors.[10]

Such practical jokes resemble those made at the expense of people's avatars in the online virtual world Second Life when users were "afk," or "away from keyboard" (Boellstorff 2008:106). Whenever a resident left his or her computer without logging off, his or her avatar "remained there, standing and looking around, sitting on a sofa, or dancing at a club thanks to an automated animation. After about three minutes the avatar's head would bow down and the word 'away' appear over it" (107). Afk is "fundamental to cybersociality"

(110), constantly reminding residents of the work of mediation and of the material processes that are at the basis of virtual worlds. In Second Life, afk was so fundamental to residents' experience that many of them used it to produce comic effects: "At a fishing dock, an afk resident's friends might push the resident's avatar off the dock into the water (causing no lasting harm, but leaving the avatar submerged among the fish). Another common practical joke was to arrange objects and other persons around an afk avatar so that it appeared the avatar was having sex, vomiting or getting ready to stab someone with a knife" (111). Users approached avatars' irreducible and recalcitrant materiality as a constitutive feature of the virtual world, a foundation for new practices and forms of attachments with performing objects rather than as a feature that needed to be masked, ignored, or repaired.

Furthermore, "in many actual-world cultures, to have someone stare at you blankly for two minutes when you approach them would be interpreted as rudeness or intoxication" (Boellstorff 2008:106). Indeed, it would be an uncanny phenomenon akin to approaching an unresponsive individual and consequently doubting whether that person is dead or alive (Jentsch 1997). But because afk was part of the fabric of daily experience in Second Life, many users displayed tolerance for "the limited capacity for avatar facial expression . . . [and] delayed or unexpected responses" (Boellstorff 2008:147). Note that "limited capacity for . . . facial expression" is an example of what Garfinkel called "deprivation of secondary information," which, he predicted, would likely lead to "states of zero-order information" and to the dissolution of the meaningful world. In communication theory, secondary information is "information that a receiver gets from a message that tells him how his own message work had been received by the other person" (Garfinkel 2008:270). Garfinkel notes that "anyone who has a student come in seeking redress for a poor exam grade, by assuming a perfectly blank stare and interpreting and answering his questions literally and in strict accordance with public protocol, will witness some intriguing effects of such deprivation" (258). In response to such deprivation, the individual will attempt to "effect a fresh normalizing"; if he is unsuccessful in doing so, the result will be "confusion: anxiety, withdrawal, compulsive repetition, over-determined reactions" (260). In contrast, residents of Second Life have continued to engage meaningfully with this virtual world not by misrecognizing or repairing or becoming anxious about the numerous technical glitches in it, which make the work of mediation that supports this world salient, but by approaching them as this world's inescapable noisy condition of possibility, which can provide the basis for new forms of aesthetic pleasure, attachment, and sociality (Boellstorff 2008:101–106).

Taking a Robotic Moment

The uncanny, according to its scholars, revolves around "the fundamental propensity of the familiar to turn on its owners, suddenly to become defamiliarized, derealized, as if in a dream" (Vidler 1996:7). It does so because "social and technical projects of immediation"—which "often successfully dampen experiential recognition of the medial and formal dimensions of media" (and, I would add, of semiotic forms in general)—"also remain perpetually unfulfilled" and fragile (Boyer 2007:59; see also Wilf 2014a:115–138; Wilf 2019:124–146). But this is the case not just because those media and semiotic forms tend by nature to be noisy and thus to reflexively point to themselves. It is also because the people who interact with them, owing to those individuals' culturally specific dispositions and predilections, focus their attention on the self-reflexive noise that media and semiotic forms occasionally produce. In the modern West, such dispositions and predilections are informed by the intellectual history I discussed above. They emerged as a reaction to Enlightenment-era semiotic ideologies that have attempted to suppress this materiality, and they account for the fact that my interlocutors in the lab often seemed more enlivened by focusing on, and even encouraging, the recalcitrant materiality and interactional incompetence of the machines they worked on than by the prospect of dampening their experiential recognition by "repair" in order to sustain the illusion of interacting with intentional entities. When Syrus's mallets failed to hit the marimba bars and suddenly unmasked its noisy "motor music," they listened carefully.

Anthropologists should do this too. If so far they haven't always done so successfully, it is because the foundations of key strands of anthropological research on communication and interaction are anchored in World War II–era technoscientific projects such as cybernetics and the design of information technologies, which had a conservative bias in that cybernetics focused on homeostasis in self-regulating systems and the design of information technologies focused on the reproducibility of signals.[11] Some of these intellectual-historical links are by now well-known. The Macy Conferences on cybernetics, which key anthropological figures such as Gregory Bateson and Margaret Mead attended (Novak 2013:150), are one example. Another example is the impact of information theory on the theories of meaning developed by Roman Jakobson and Claude Lévi-Strauss (Geoghegan 2011).

An additional, relatively unknown intellectual-historical link that has skewed the study of human interaction and human-machine interaction are the cybernetic roots of Garfinkel's ethnomethodological theory. This intellectual history has informed the study of human-machine interaction in the form of the idea that humans are inclined to maintain interactional homeostasis when interacting with machines by repairing the gaps in machines' rudimentary behavior and by compensating for machines' inability to engage in repair. In contrast, interactional noise is often a focus of attention, an object of fascination, and a source of aesthetic pleasure that people strive to prolong and even deliberately generate when interacting with machines. As interaction with different kinds of machines, performing objects, and technologies of mediation becomes a key dimension of sociality in the modern West, a more complex understanding of such interaction is needed to better account for this sociality.

I ended the previous chapter with the observation that Syrus's embodied features produce a distinct style of playing that serves as a reminder that fantasies about complete stylistic flexibility and malleability will always be limited or curbed by the infrastructure required to realize them. By itself, Syrus's residual style already problematizes the argument, which I discussed in chapter 2, that companionship with a sociable robot spares us the need "to tolerate disappointment and ambiguity" and to "accept others in their complexity," and that "the first thing missing if you take a robot as a companion is *alterity*, the ability to see the world through the eyes of another. . . . If they can give the appearance of aliveness and yet not disappoint, relational artifacts such as sociable robots open new possibilities for narcissistic experience" (Turkle 2011:55–56). Syrus's residual playing style was a recalcitrant feature that team members had to accept and tolerate even though it frustrated their research objectives. Furthermore, rather than merely tolerating this feature, team members occasionally turned it into an object of humorous exploration, as Matt's joking imitation of this style with the electric piano and Kim's exclamation—"Syrus music!"—demonstrate. As I showed in the present chapter, during my fieldwork in James's lab team members frequently displayed this stance not only in relation to some of the recalcitrant features that Syrus displayed as part of its normal functionality (such as its distinct style of playing), but also in relation to many of its malfunctions. Furthermore, at times they even actively helped Syrus deviate from the path they had intended it to follow or willingly demonstrated the artifice behind the very idea that Syrus could follow that path to begin with.

In addition, team members' stance toward Syrus's recalcitrant materiality suggests that, far from opening "new possibilities for narcissistic experience" (Turkle 2011:56), in which one does not need to "accept others in their complexity" and "alterity" (55), taking a sociable robot as a companion, and certainly developing one, can provide the basis for a renewed appreciation for humans' complexity and alterity, too. Parker, a faculty member in his mid-thirties whose specialty is the use of artificial intelligence in music composition and who has collaborated with James on different aspects of Syrus's robotic musicianship, elaborated on this point at length when I interviewed him in his office:

> The whole nature of being human or even a primate or a mammal—we are highly creative. We are always adapting to new scenarios in ways which do not fall into a clear set of rules. And you are forced to recognize this when you try to design artificial systems. You realize that even simple things require tremendous creativity. I try to make people realize that the simplest and most automatic things we do are often the most complex things we do. Take social interaction as an example, since interaction is a big dimension of what we had hoped we could teach Syrus to do. People used to think it's a simple thing, and now they realize that it's one of the most complex things. The computation that is required for that is enormously complex. There are these crazy regressions: "This person is thinking, thinking of me thinking of him," and so forth. And you get these insane modeling problems. And we humans are doing it all the time. It's something that we take for granted, but just because we all do it doesn't mean it's simple. In order to interact you have to have a reliable estimate of what's going on around you. Our techniques, even if they are 90 percent good, are not enough. Computers still have a very hazy view of what's happening around them. Or take another issue. When you are predicting what's happening next in an ensemble, you are using a tremendous amount of information. Just think about the visual aspect, which we also try to teach Syrus. But even if you forget about this aspect, you know, we bring with us all the types of music information and we use all of that—the melody, the rhythm—to predict everything. We use all kinds of correlations between the different dimensions. We don't just use the melody. We use all kinds of contextual information to anticipate what's happening next. And it's very hard to model that. And for me one of the big gaps is that Syrus is not aware of the relationship between things. It has some idea of what's happening locally, but the most relevant piece of information is not what happens locally but what happened a bar ago or like four bars ago and what will happen next. These are all things that are simple to state but hard to program.

Parker describes the different challenges that the team struggled with as a result of the fact that they attempted to design Syrus to be able to interact with human musicians in a flexible way that exhibits context sensitivity oriented both to the past and to the future—that is, beyond Syrus's immediate present—by equipping it with an embodied form, in addition to other strategies. Crucially, this attempt led the research team to appreciate the complexity of even the simplest tasks that humans perform on a daily basis.

Matt, too, argued that the fact that the research team had only managed to train Syrus to engage in a very rudimentary form of musical interaction with human musicians was valuable because this experience taught them about the complexity of human interaction, especially in relation to its embodied dimensions:

> We did a lot of research about interaction, like the anticipation and synchronization between players through bodily movements. We learned a lot about how performers perform with each other by studying how we perform with a robot, also the embodiment and studying what that means for a performer and how that affects how we play. I think all of that is a really important research that has come out of that. And that's what's important, not how much we managed to turn Syrus into a bulletproof commercial product or anything like that. It has become more about using Syrus as a platform for studying various aspects of musical interaction and different technologies.

James, who had hoped that Syrus's embodied features would play a crucial role in its ability to inspire and "activate" its human interlocutors, similarly commented:

> As I told you, I worked for many years with generative algorithms but I wanted the computer to inspire me and I missed some things like the acoustic richness and the visual cues. And it's funny, most people who see Syrus think that it has mechanical skills that surpass those of humans, that it can play faster and more things. Now, there are some things it can do that people cannot do. If you have four arms and eight mallets, you can play complicated chords. But in terms of virtuosity in general, a mediocre player can do many more things that Syrus cannot do in terms of speed, types of runs, and tightness with other musicians. The moment it has to switch from one place to another on the marimba at a certain speed, the moment there is a delay in the mechanical system—you know, this is not a robot that builds cars and that can do so faster and better than a human. This is a very different kind of robot. And what I also learned in relation to that, especially about inspiration and creativity, is that it is not enough to have a statistical model even if it's done expertly. It doesn't really capture the essence, the *je ne sais quoi*. We don't have a model of aesthetic judgment, what people like and don't like, a cognitive model. So barring that, we have to make do with simplistic models. But it doesn't really work because whatever we develop doesn't have an inner ear that can distinguish what is good and what is not good. So we try to use proxies. But at the end of the day, innovation is so subtle—you just felt that something felt right. And you realize, when you try to program all this, how subtle creativity is, how many decisions are involved when we act creatively. The search base is vaster than the particles in the universe. There are just too many possibilities. So, yes, some of it we can model with algorithms and stuff, but the people we consider great, they are able to take that and something exceptional emerges. Those insights are very hard—you can make post hoc rationalizations, etc., but it's

not enough. There is something different about art, things that are about the human experience that we do not know now how to program.

As all of these comments suggest, by trying to develop Syrus as a fully embodied robotic musician that can participate in a real-time joint improvisation with human musicians, the research team developed a renewed appreciation for the complexity of human beings' creative agency. Indeed, if anything, of the two jazz-improvising computerized systems that I have discussed thus far, it is David's system, not James's, that comes closest to Turkle's prediction about the "new possibilities for narcissistic experience" that sociality with relational artifacts might open up (Turkle 2011:56). However, this is not in spite of but rather precisely because David consciously avoided what Turkle calls "the robotic moment" (Turkle 2011:55), that is, the contemporary turn to sociable robots as companions. David deliberately chose to develop a software program that generates sounds by means of a simple tone generator rather than to develop a fully embodied robotic musician. Because David chose not to give his system an embodied robotic form, it was easier for him to feel that his artificial world functioned as a suitable replacement for the musicians with whom he was so frustrated. Trying to design and interact with a fully embodied robotic musician would have forced him, as it forced James's research team, to come to terms with the recalcitrant nature of the system's materiality and to gain a renewed appreciation for the complexity of human musicians' creative agency. By limiting himself to interacting with a software program that is always already more abstract and context insensitive—and, in addition, by limiting the program's interaction capabilities to very few dimensions of human sociality—David minimized the risk that he would be disappointed while interacting with his system. Thus, avoiding rather than embracing what Turkle calls "the robotic moment" led David to experience the kind of self-focused relationality that Turkle associates with the rise of robotic companions.

Furthermore, it is significant that even though David realized it was impossible for him to program an aesthetic fitness function in the context of the genetic-algorithms computational framework, this realization did not lead him to James's conclusion about the irreducibility or singularity of human creativity, but to the opposite conclusion. During a training session, David told me:

> My system is fairly modest in what it does. All it has to do is improvise single-note solos over a given chord progression. To some extent it's playing fastball because it is playing straight-up jazz to a fixed chord progression, and it's gonna hit the harmonic changes and right there that gives you a deep

structure, just if you follow the chord changes. The rhythmic structure is set. It's always going to play a fixed tempo. And its architecture was designed to be very fast and very robust and very responsive and not to think hard about what it does. My whole philosophy in building any kind of software is to start from simple and only get complex if you run out of simple. And I never really ran out of simple. I can add more interesting things to it but in terms of its architecture, it's just boneheaded simple. And this is why I think it puts into question notions of creativity. . . . This might sound cynical . . . but I've never felt that creativity, whatever the hell it is, is inherently human. It's just a thing. It happens. If it happens to happen from a program—OK.

Although David acknowledges that he simplified the problem or the search space that his system needs to deal with, he argues that the system's ability to successfully deal with this simplified problem or space means that human creativity may be simulated by computerized systems. Thus, because David designed a relatively simple system that could function well in the context of a very restricted dimension of human creativity, he came to think of human creativity as simple, too. His notion of human creativity came to be informed by the simplicity of his system and of the tasks that this system was designed to perform. In contrast, because James and his research team tried to design a much more complex system that could function in more and more complex dimensions of human creativity, their notions of human creativity became saturated with the complexity of their system and of the insurmountable tasks this system was designed to perform.

These insights should make it clear that, from an anthropological perspective, whether or not new technologies and computational models of machine learning are already available or will soon be available by means of which the different challenges that Parker, Matt, and James described might be addressed is beside the point. Rather, as I argued in the introduction, inasmuch as the technologies that human beings create are socially embedded, they will always be informed by (rather than supplant) the socially and culturally specific contexts that shape the motivations for their design, their architecture, humans' experience of interacting with them, and humans' evaluation of their output in relation to the problems they were designed to solve. Conversely, such technologies will impact and lead to changes in those socially and culturally specific contexts.

As I now turn to show in relation to the field of computer-generated poetry, the nature and intensity of such bidirectional influences often vary depending on the precise ways in which digital computation is mobilized for the purpose of making art. In the field of computer-generated poetry, code's affordances and limitations have a much bigger impact on the poet's understanding of his or her creative self, practice, oeuvre, and aesthetic goals because the poet's creative agency is intertwined with code in a much more significant way than in the jazz-focused context I explored here and in previous chapters. Consequently, this second ethnographic context raises a different set of questions, questions that turn on the relationship between computational indeterminacy, poetic potentiality, and creative intentionality.

PART II

Poetry
Indeterminacy, Potentiality, Intentionality

5

Computer-Generated Poetry and Some of Its Aesthetic and Technical Dimensions

"It Should Be Python 3!"

My path to the weeklong computer-generated poetry workshop that is the focus of the next four chapters was rather convoluted. The workshop, which took place in New York City in a housing and working complex dedicated to the arts, was organized by a nonprofit organization founded by a group of people who are interested in poetic computation. I first became aware of this organization in the summer of 2013, when I met by chance and then interviewed one of its founding members in what was then the organization's location in Brooklyn. Although I had intended for a while to attend the organization's workshops and classes, the opportunity had never arisen. However, a few years later I learned that the organization would offer a workshop dedicated to computer-generated poetry at a time when I would finally be available. I submitted an application in which I described my research interests in the use of computation for creative purposes. I also mentioned that I am a published poet. After a few weeks I received an email from Jason, an established writer of computer-generated poetry in his early forties, who was one of the workshop's key facilitators. He notified me that I was accepted to the workshop. I quickly made my travel arrangements and was soon on my way to New York.

The workshop took place in a large room with a big window that faced an inner courtyard. At the center of the room stood two large tables, around which the other fourteen workshop participants, who were mostly in their twenties and early thirties and who had some familiarity with coding, sat with their laptops. The facilitators sat in the front of the room next to a media unit that consisted of a large television screen on which they could show the participants what they were doing with their laptops. On a side wall stood a cabinet full of books that focused on the technical, aesthetic, and social dimensions of computation. An adjacent room functioned as a performance and exhibition space in which the projects made by participants in the organization's past workshops and classes were displayed.

On the workshop's second day, Shiv, a participant in his early thirties and a master of fine arts student in digital art and new media, presented a poetry generator that he had created based on a prompt one of the facilitators had given to the participants in the previous session. The prompt required the participants to write a program that generates "a lot of variation." After hooking his computer to the media unit, Shiv ran the program he had written. The following output text was projected on the media unit's screen:

> ilehhiloilehhiloilehhiloohoelodeohoelode
> odillodeodillodeodillodeeleidehieleidehi
> odillodeodillodeodillodeeleidehieleidehi
> odillodeododdehiehoddodeeleidehoidiohilo

Shiv proceeded to read the output text, struggling with the pronunciation. After he finished reading, Jason said: "OK, so this is a—it doesn't matter what these words are. There is language, in that there is a list of words. There is a square of it and it's pleasing to look at. And, in fact, it is potentially pronounceable. There might be ways that you can say it. Other thoughts?" Carl, a participant in his late twenties who works in an apparel start-up, said: "Regarding the pronounceability, a text-to-speech [software] might be able to handle it although there is something interesting in seeing you grappling with where a word starts and ends, things like 'hi' and 'hello,' these very guttural sounds. A text-to-speech might do something way different." Trish, a digital media artist in her late twenties, asked: "Can you try it?" Jason said: "Well, if you are going to use Google, why don't you throw it at Google Translate and see what language it identifies it as, what language model it selects it as?" As he was saying this, he laughed. "It should be Python 3," he added. "That's the answer!" At that everyone in the room burst out laughing.

On the surface, Jason's joking suggestion to let Google Translate determine the language in which the output text is written, as well as that the correct answer is Python 3, merely points to the distinction and connection between two key components of computer-generated poetry, namely a computer program written in a programming language such as Python 3 and the output text that is produced when the program is executed. However, it is more deeply motivated than that. To begin, when Google Translate "detects" (as its interface puts it) a language, it reveals the deep linguistic structure that is responsible for whatever surface-level stretch of text the user asks the software to translate. Jason's suggestion thus pointed to a hierarchy of sorts between

Python 3, as a kind of deep structure, and the resulting output texts of the program written in this programming language, as a kind of epiphenomenal surface-level structure.[1] At the same time, Google Translate can detect a text as one that is written in a specific language because there is a clear and easily detectable connection between the text and the language (e.g., a connection that is based on an alphabet or a vocabulary). In contrast, it is impossible to identify the programming language in which Shiv wrote his poetry generator solely on the basis of this generator's output text. This would be the case even if the output text were not nonsensical. This discrepancy provided the comic element of Jason's suggestion. Indeed, when Shiv followed Jason's suggestion and "threw" his text into Google Translate, as Jason put it, Google Translate "detected" that the poem was in Finnish, to everyone's delight.[2] Jason used Shiv's poem as an opportunity to point both to the causal and hierarchical relation that exists between a poetry generator's code and its output texts and to the space of indeterminacy that separates the two. Jason's joking suggestion thus touched upon a number of fundamental dimensions of computer-generated poetry as a specific artistic genre.

The present chapter describes some of these dimensions and how the workshop facilitators attempted to socialize the workshop participants into them. The next three chapters focus on the potentialities and tensions that emerged around those dimensions during the workshop, and on their cultural and intellectual historical sources.

The Primacy of Code and Computation

Computer-generated poetry is first and foremost a type of "digital literature," that is, "literary work that requires the digital computation performed by laptops, desktops, servers, cellphones, game consoles, interactive environment controllers, or any of the other computers that surround us" (Wardrip-Fruin 2010:29; see also Tomasula 2012:483). In all of these cases, the "computer cannot be seen only as a tool for writing, but rather as a partner in creative processes" (Koskimaa 2010:129). More specifically, computer-generated poetry is a subgenre of digital poetry, which "refers to creative, experimental, playful and also critical language art involving programming, multimedia, animation, interactivity, and net communication" (Block, Heibach, and Wenz 2004:13). Although the workshop in which I conducted fieldwork touched upon different kinds of digital poetry, such as time-based, kinetic digital works that change as the viewer engages with them, its main focus was computer-generated poetry, or "works of text generation" that typically "transform or reorder one set of base texts of language (word lists, syllables, or preexisting texts) into another form" (Funkhouser 2007:36). Such "text generators" can "rapidly produce many poems" by "using programmatic formula that selects words from a database to create output" (79). Algorithmically generated poems can be based on different permutational procedures: "Works are either permutational (recombining elements into new words or variations), combinatoric (using limited, preset word lists in controlled or random combinations), or slotted into syntactic templates (also combinatoric but within grammatical frames to create an image of 'sense')" (36; see also Gervas 2019; Saemmer 2010:164–165). These procedures are not mutually exclusive. Historically, this type of computer-generated poetry marked the beginning of digital poetry at large, and many of its aesthetic principles and sensibilities have continued to inform the cultural order of digital poetry in general (Funkhouser 2007:6–7, 41). This is why the workshop facilitators chose to introduce the participants to digital poetry by organizing the workshop around a number of classic computer-generated poems written in the late 1950s and 1960s.

In writing computer-generated poetry, a poet writes code whose execution by the computer results in output texts. The primacy of code affects the poet's engagement with this art form from the outset of his or her creative process. For example, when I asked Jason to describe how he comes up with ideas for new computer-generated poems, he thought for a few seconds before responding:

> Well, every year I do a New Year's poem, which is something that Georges Perec would do every year. So it's an obligatory occasional poem. And I might be thinking about doing a one-line basic program for Commodore 64. Or one year I did, I asked, can I write a self-contained program that is 256 characters long? I wanted it to be a poetry generator, not just something that produces language or produces pronounceable text. So the idea was writing a core program in 256 characters. That was the initial concept, the same way someone might say, "I think I'll write a sonnet."

Jason's response is noteworthy for two reasons. First, it points both to the continuity and discontinuity between computer-generated poetry and previous forms of poetry. Deciding to create a poetry generator that consists of a program made up of 256 characters is similar to deciding to write a sonnet, in that both are decisions to write in the framework of specific constraints. However, in contrast to earlier forms of poetry, a key dimension of the craft of the poet who decides to write a computer-generated poem is inventing constraints by means of code rather than using existing constraints. Jason expresses this by comparing his own poetic practice with that of Georges Perec, who was perhaps the most famous and prolific member of the Oulipo—a literary group founded in 1960 whose core focus is writing literature based on constraints that are mostly (though not necessarily) self-fashioned and mathematically informed.[3] Second, Jason's response to my question points to the transformation of code and computation into fundamental dimensions of the poet's creative agency even prior to the actual writing phase, that is, when the poet might think of an idea for a new computer-generated poem. The poet might begin the creative process with the idea of writing a specific type of program (e.g., "a self-contained program that is 256 characters long") and/or by means of a specific type of computational hardware (e.g., a Commodore 64).

In an equally consequential way, the practical and normative orientation to code and computation as fundamental dimensions of the poet's creative agency might find expression in the poet's inclination to observe the world through a kind of filter that translates this world into phenomena and categories that are amenable to computational representation and manipulation. Chris, an experienced writer and researcher of different forms of digital poetry and another of the workshop facilitators, described to me a specific computer-generated poem that he had written a few years prior to the workshop and that he presented in it:

> I was carrying around a notebook and I was in the Los Angeles area and I was just thinking a lot about cars, the ways that naming, trade names are assigned and devised for cars. And I thought, well, maybe I'll just find out—like, let me see what sorts of categories, what sorts of clusters I could put these car names into. And it was interesting to me because a lot of cars are named after horses, which were the former mode of transportation. . . . And then there are cars that are named after Native people in the United States. You have Dakota, Cherokee, Comanche. And then there are certain animals. They sound silly if they are the wrong animal. Like you can have fast birds—they can't be ostrich—it has to be fast birds like the skylark, or fast animals like the lynx or the jaguar. So I started thinking about these things in terms of these categories first, and then I came to think about—one of the things I'm often telling people is that in English nouns and verbs can be the same word. . . . So I had this concept and I started thinking about language and about the way that corporations and people involved in these companies have assigned these names, and I started thinking about the American context, you know, Los Angeles, and then I realized that it would be possible to just have one utopia or dystopia: it would be a city that's all cars, that has no people at all. And I thought, well, you can have this work made of language where all the language is the names of cars. Nothing else. And that is what was developed. And then I made a piece in Python.

Chris's account of his poem's origin shows how his perception of the world is informed by or filtered through the possibility of its computational representation and manipulation. His reflection on the topic of the model names of cars takes the form of trying to sort them into "categories" and "clusters." This sorting leads him to identify a number of key categories and clusters, such as specific types of animals and groups of people. In the process, he makes an observation about grammatical categories, namely that in English "nouns and verbs can be the same word." This thought process leads him to realize that he can represent a city populated only by cars through "work made of language where all the language is the names of cars." The realization of this idea in a computer program that is written, in this case, in Python, is a natural next step, because the thought process that led to this idea was always already computational in essence—that is, it was informed by what a computer program can represent and explore.

Chris's description points to the phenomenological manifestation of what Lev Manovich describes as "the projection of the ontology of the computer onto culture itself," which he has analyzed with respect to computer programming:

> Computer programming encapsulates the world according to its own logic. The world is reduced to two kinds of software objects that are complementary to each other—data structures and algorithms. Any process or task is reduced to an algorithm, a final sequence of simple operations that a computer can execute to accomplish a given task. And any object in the world—be it the population of a city, or the weather over the course of a century . . .—is modeled as a data structure, that is, data organized in a particular way for efficient search and retrieval. (Manovich 2001:223)

Chris's account suggests that the logic of computer programming can become the cognitive filter through which the writer of computer-generated poetry might observe and organize the world.

During my conversation with Jason, I learned that his understanding of whether he has a unique poetic style and what this style consists of is similarly informed by code and computation. A few minutes after he mentioned Georges Perec, we had the following exchange:

EITAN: Do you have a unique style?
JASON: Oh, sure, sure.
EITAN: How would you describe it?
JASON: Well, style can arise—if you adopt the sort of constraints and writing techniques that the Oulipo has developed, then although that cer-

tainly doesn't eradicate one's writing style, it's quite influential on it. If you choose to write a univocalic poem, for instance, then, depending upon which vowel you use, if you write a poem and use only the vowel U, it's going to be very guttural. The things that are available to you there will be very specific.

EITAN: What kind of constraints do you tend to gravitate toward?

JASON: Well, I'm interested in very difficult constraints that are on the edge of possibility. So a lot of my work, part of it, for me, is the issue of setting up a challenge and then creating something—where I would like for the text to sound natural when it's challenging or difficult to do that. Now, what natural means is different in different contexts. Sometimes I am just interested in pronounceability or recognizability as English. Sometimes I am more interested in modeling the style of an existing author like Samuel Beckett. But regardless of that, I think my body of work is identifiable because a lot of what I do is writing short programs that are self-contained.

Jason responded to my question by discussing primarily the type of programs he tends to write ("short programs that are self-contained") and the kind of constraints these programs embody ("difficult constraints that are on the edge of possibility"). Although he writes such programs and fashions such constraints for different purposes (e.g., making a text sound natural and recognizable as English, or modeling the style of a well-known author), the kind of programs and constraints that he writes and fashions have specific features. As far as he is concerned, these code- and computational-level distinguishing features define his unique poetic style and represent an important dimension of his artistic identity.

The fact that code is the basis of computer-generated poetry leads poets to relate to the poetry generators they and others create in terms that disrupt widespread notions about the quintessential work of art as a highly original work epitomizing singular perfection that cannot be improved upon. What affords this disruption is the combination of a number of factors. First, the code of existing poetry generators is easily accessible because poets frequently make their code available to the wider public. Second, it can be easily modified. Third, even small modifications in the code can result in significantly different (and, to writers of computer-generated poetry, sometimes more aesthetically pleasing) output texts. Rather than a locus of distant uniqueness, as Walter Benjamin described the work of art, using notions such as aura and singularity (1969:222–223), a poetry generator becomes a locus of proximal multiplicity, as it were, in that every practitioner of computer-generated poetry can both gain access to an existing generator's code more or less easily and change it in many different and consequential ways. Fourth, inasmuch as many poetry generators regenerate themselves and can result in numerous different output texts, the aesthetic principles that inform them focus not on "creating singular artifacts" to begin with, but on creating "manifestations" (Funkhouser 2007:13)—dynamic processes or texts that aim "to be infinite, not eternal" (83). Such practical and aesthetic principles disrupt the notion of textual or artifactual singularity and uniqueness in time and in space, which is one of the cornerstones of Romantic ideologies of creativity.

The poets I worked with embraced rather than bemoaned the fact that the poetry generators they create can be easily modified by other people, who thereby might create different and even more interesting versions of those generators. They approached this fact as a way to encourage other people to engage with their poetry. In our conversation, Chris explained:

> So, for me, creating a computer program that embodies as much of my process as possible is important because I want people to study, modify, remix, and work with that program. That's the case with one of the poems I wrote, where, I mean, it's been given as a classroom assignment in many places. Hundreds of people have made their own versions of this poem and there are more than thirty better versions published by different authors and poets that are online that I know about. And if I had just written a program that produced some output text and I edited down only the output text and published it [and not the program], none of that would have happened. So I want to open the process to people in that way.

Chris argues that had he published only an output text of his poetry generator—an act that would be equivalent to publishing a "traditional" poem—he would have made it more difficult for other poets to engage with his generator. What could have been an object of anxiety for Chris, had he adhered to the Romantic notion of the artwork as a locus of artistic singularity—namely, the fact that other people can modify his poetry generator and make its output texts aesthetically "better"—instead becomes an object of pride and satisfaction. The number of people who become invested in one's computer-generated poem to such a degree that they are driven to experiment with and alter its code becomes a source of satisfaction even—and especially—if such modifications improve one's poem.

Socialized into Computational Indeterminacy

The primacy of code and computation in computer-generated poetry has a number of other implications for the poets who are focused on writing this form of poetry. First and foremost, many of my interlocutors experience computer-generated poetry as an art form that is based on a fundamental dimension of computational indeterminacy of both process and outcomes—a feature I will discuss in greater detail in the next chapter. To begin, the coded instructions whose subsequent execution by the computer results in output texts are relatively abstract (written, as was the case in the workshop, in a programming language such as Python 3). As a result, neither the instructions' writers nor their readers can immediately and fully comprehend their meaning and implications. In addition, many poetry generators incorporate randomness into their architecture by means of the programming language's random function or by virtue of the reader's input via an interface, while other poetry generators are designed to produce deterministic yet seemingly random sequences of words. As a result, the exact nature of their output texts cannot be anticipated in advance with complete accuracy, even when it is possible to figure out to some degree what their code aims to achieve. Therefore, poets must cultivate a nuanced understanding of the relationship between the code they write and its potential output texts, and especially of how making specific changes in the former can impact the latter.

During the workshop participants were socialized into, and given the opportunity to cultivate, this understanding in two ways. First, the facilitators gave the participants the chance to experiment with the code of a number of classic computer-generated poems. By making changes in a poem's code and observing their impact, participants were exposed to and internalized the nuances of the relationship between the two. For example, in the workshop's first meeting, Jason screened with his laptop a version of the code of Theo Lutz's classic 1959 computer-generated poem, *Stochastic Texts*, which uses computer-based randomness to produce output texts by taking as its source material key words from Franz Kafka's story *The Castle* and slotting them in different sentence structures. Lutz's poem holds an important place in the history of digital poetry: it marked the beginning of "the pursuit of composing poetry by using computer operations" (Funkhouser 2007:37). Here is a stretch of an output text of this poem (translated from the German):

> NOT EVERY LOOK IS NEAR. NO VILLAGE IS LATE.
> A CASTLE IS FREE AND EVERY FARMER IS FAR.
> EVERY STRANGER IS FAR. A DAY IS LATE.
> EVERY HOUSE IS DARK. AN EYE IS DEEP.
> NOT EVERY CASTLE IS OLD. EVERY DAY IS OLD.
> NOT EVERY GUEST IS ANGRY: A CHURCH IS NARROW.
> NO HOUSE IS OPEN AND NOT EVERY CHURCH IS SILENT.
> NOT EVERY EYE IS ANGRY. NO LOOK IS NEW.

> EVERY WAY IS NEAR. NOT EVERY CASTLE IS QUIET.
> NO TABLE IS NARROW AND EVERY TOWER IS NEW.
> EVERY FARMER IS FREE. EVERY FARMER IS NEAR.
> NO WAY IS GOOD OR NOT EVERY COUNT IS OPEN.
> NOT EVERY DAY IS LARGE. EVERY HOUSE IS SILENT.
> A WAY IS GOOD. NOT EVERY COUNT IS DARK.
> EVERY STRANGER IS FREE. EVERY VILLAGE IS NEW.[4]

Jason then began to modify the poem's code in real time while talking to the participants:

> What I want to show you is that it's very easy to take this classic digital poem and to make a modification which is significant. Not to make a great literary work, perhaps, but to actually make something different out of it. What we have here is a Czech author, German language, story of alienation, which is the basis for this. And I'm a good US American [Jason laughs], and I want a different story of alienation that is written in English. I am going to make this based not on Kafka's *The Castle* but on Melville's *Moby-Dick*. And so what I'll do, I'll keep all the predicates the same—"silent," "strong," "good"—that's fine, but I'm gonna change the subjects: "ship," "captain," "mast," "hammer," "harpoon," "leg," "deck," "boat," "rail," "coffin," "harpooner," "sea," "sail," "sky." [Jason edits the code as he speaks.] So, I made my changes. I am putting in singular nouns.... Ok, let's take a look at what this does. [Jason runs the program and reads the results that are displayed on the television screen:] "Every sail is free. A sail is large. Every coffin is far. A sail is silent. A hammer is near, therefore not every sail is silent. No sky is angry, no sky is large." [Jason looks at the participants.] Now, I submit that this is a different literary generator, just by changing these words, not by changing the rules.

Jason encourages the participants to experiment with Lutz's classic poem not by changing the poem's basic rules, which create specific sentence structures, but by changing the pool of words that might appear in these sentence structures. In subsequent stages of the workshop, the facilitators asked the participants to also modify the rules of this and of other classic computer-generated poems. In an email to the participants, Jason wrote: "A big point of this week is that you can get involved with code, computation, and digital poetry/art by making simple modifications to simple programs. You have been shown many [classic] programs that are available to you in Python [through links shared by the guest lecturers].... You've already dived in, and by just making small changes (undo them if something breaks and try again), you can develop your own work."

Another important pedagogical strategy that the facilitators used during the workshop moved in the opposite direction, namely from the output texts of specific poetry generators to the generators' code. Time and again, after one of the workshop participants read an output text of a poetry generator that he or she created, the facilitators would ask the participants to try to decipher the underlying computational logic of the generator's code based on that output text. For example, when Adrian, a participant in his early thirties who works for a financial start-up, finished reading the output text of his generator and then wanted to describe the generator's code, Jason stopped him and encouraged the participants to comment on the output text. When one of them offered what she liked about it, Jason stopped her too and said:

> Yes, but what do you see there? What's happening there in the code? What do you imagine is happening here? To be readers of this—it's nice to see the output text, and I want everyone to respond to this output text first because I want to know what people think of it. But, really, to be readers of this kind of work means that we have to read the code as well in many cases. And when we don't have the code then we have to be able to read the process, to figure out how it's [the output text] being produced.

As the workshop progressed, participants cultivated a better understanding of the relationship between the code they wrote and its potential output texts, and of the relationship between the output texts of poetry generators created by someone else and these generators' underlying computational logic.

Learning to Inhabit a Text-Distributional Aesthetics

On the workshop's second day, after a few of the participants had commented on their burgeoning understanding of the relationship between code and the output texts that are produced when the code is executed, Jason turned to the seated participants and said: "This is a new engagement with writing. What we have is a distribution of text rather than a text that comes out of systems like this. So we need to be creators not of texts but of text distributions. We need to be this kind of writers." In a conversation I had with Jason in a coffee shop, I asked him to clarify what he meant when he advised the participants to start thinking of themselves as "creators not of texts but of text distributions." He leaned toward me so I could hear him amid the loud conversations around us and said:

> I can treat the output of many of the classic text generators we worked on during the workshop either as a single generated text or I can treat it as a distribution from which I can draw a text. So those are just two different ways of approaching the same underlying process. It's like, for instance, if I went to a country, let's say Lichtenstein, and I wanted to know who the people of

Lichtenstein are. So one way I could find out is like a Gallup poll. I could sample people from the population. I could say, "OK, let me get a dozen of random people and I'll talk to them and I'll learn about the population of Lichtenstein." Or I could get like a giant gymnasium and I could get everyone in Lichtenstein there [Jason starts laughing] and then I could see: "Oh, there it is." But the point is, however you choose to think of the process—either sampling from a distribution when you use randomness or [doing it] exhaustively, it doesn't really matter. A text that is exhaustive could be sampled! You could pick from those lines at random if you'd like! I could just generate those lines at random, pick one line and show it, leave it on the screen for a while, and then take another line and show it. It's up to you. That's why one of the things I would say is that randomness is not a big deal in this case. It's just a way of getting a perspective on this distribution. You could print it all or you could sample from it. . . . So, for me, when I am talking to people about a shift in thinking from writing a text to—what is digital poetry? What does it do?—I think about, first of all: can you understand the difference between defining a distribution of text as opposed to defining a text? What the artist or writer [of computer-generated poetry] is typically doing is defining a distribution that can be sampled, not defining a particular text or a particular visual art or whatever.

Jason's response exemplifies the fundamentally text-distributional aesthetics that underlies much of the computer-generated poetry on which the workshop focused. The output texts produced when the computer program the poet writes is executed are approached as samples of, or as a window into, the specific text distribution that the poet defined by means of the code he or she wrote. The core essence of such a text distribution is typically a general linguistic texture or pattern, whether such a texture or pattern is the result of code that incorporates randomness in its architecture or of code that produces a deterministic sequence of syllables or words. Lutz's *Stochastic Texts* exemplifies these poetic dimensions. Jason argues that to understand the aesthetic essence of a computer-generated poem such as this, it would be enough for the reader to look at the short stretch of this poem's output text that I quoted earlier, for example, or at any other stretch of comparative length that can be generated by executing this poem's code at different times. This would be akin to a random sampling from a distribution. There is no need to look at the immense number of possible manifestations of this text distribution, which would be tantamount to trying to get all the people of Lichtenstein into "a giant gymnasium" in order to understand the essence of this country's population.[5]

The most aesthetically important variety in the context of the kind of classic computer-generated poetry that was the focus of the workshop is thus typically (though not always) the variety that results from the differences between different computer-generated poems as distinct poetry generators, not from the differences between the different output texts or stretches thereof that can be generated by a single poetry generator (whether or not it incorporates randomness) when its code remains unchanged. The different output texts or stretches thereof that can be generated by executing such a single poetry generator's code are appraised in terms of the underlying linguistic texture that they instantiate and that unifies all of them. This underlying texture (or distribution) gives a specific poetry generator its aesthetic value, and it is in light of this texture that the poetry generator is compared to other poetry generators. This is why, as I will discuss later in this chapter, most of the people I worked with saw very little sense in indefinitely continuing to generate more output texts from the same poetry generator by executing its code time and again, even if such texts were never identical to one another because the generator's architecture incorporated computer-based randomness.[6] They felt that they understood fairly quickly the "gist" of what the code can do—that is, the general linguistic texture it can produce—and they consequently experienced the results of its continued execution as more of the same.

Later in our conversation, when Jason and I discussed the absence of narrative development in many of the computer-generated poems—both classic and recent—that were analyzed in the workshop, he clarified these poems' aesthetic value in relation to the notions of text distribution or linguistic texture in this way: "If you're looking at a river, it's neither surprising nor suspenseful. But it still can be nice to look at a river, right? Now, you could say, 'Hmm, this idea of sampling and this idea of a texture of language, something that's more like a river and less like a narrative event, a reflection point, surprise or suspense—maybe I don't want to do this distributional thing.' But you could also say, 'Maybe there are distribution-like things.'"[7] In a 2021 podcast, the musician Brian Eno described his ambient music in identical terms:

> I came up with this piece called *Neroli* [1993], which is an hour-long piece and which at the time was the longest I could make it. And I started thinking then about music that doesn't have a beginning or an end, that is just theoretically infinite. Well, that didn't become possible until the 90s when I started working with computers and it was possible to make the kind of programs I make now, where the music effectively never repeats. And my ambition then was always to try to make an experience a little bit like sitting by a river. So you're sitting by the river, it's always the same river but, as you know, it's never the same river twice. So every time you look up at the river it's doing something a little bit different. Now it's not like watching a film where suddenly the river turns blood red or gets much bigger or something like that. I didn't want drama. I

just wanted something like nature . . . subtle variations . . . that stay within a kind of range of possibilities, and [to] explore that range rather randomly. (Rubin 2021:42:07–43:30)

Jason mentioned Eno a couple of times during the workshop. It is likely that he modeled his response to my question after Eno's description of his ambient music, which Eno had probably provided prior to time of the workshop in some other context that Jason was aware of. Eno's description not only highlights the aesthetic sensibility that informs the creation of "distribution-like things" that have no beginning, end, or narrative-like development, to quote Jason; it also points to the availability of digital computers and of computer-based randomization as important conditions of possibility for the realization of this sensibility.[8]

During the workshop, participants were encouraged to inhabit this kind of text-distributional aesthetics time and again. For example, in the workshop's first session, Jason instructed the participants to modify the code of one of the classic computer-generated poems that he presented earlier, which, in addition to *Stochastic Texts*, included Christopher Strachey's *Love Letters* (1952) and Alison Knowles and James Tenney's *A House of Dust* (1967).[9] After fifteen minutes he asked Laura, a participant in her early thirties who works at an investment bank as a financial analyst, to read the results of her modification. Laura, who chose to modify the code of Lutz's *Stochastic Texts*, looked at the screen of her laptop and read: "No cage is empty. Every house is full." She then stopped and looked at Jason, who then said emphatically: "More than that! You just gave us a fortune cookie. We want to know a solid of what's happening!" Jason used the term "fortune cookie" to argue that a onetime textual event, such as a single sentence, does not provide a large enough sample on the basis of which one could infer the linguistic texture produced by a poetry generator. To discern such a linguistic texture or pattern, a larger number of events is needed. For example, after Jason demonstrated to the students how to modify the code of *Stochastic Texts*, he said: "Now, I'm going to put it into a loop. I want to see ten of these [sentences] at once instead of one. Because I want to get an idea of what the texture [produced by the modified poetry generator] is."

In a prior exchange, during which Jason discussed *Stochastic Texts* for the first time, he explicitly related the notion of a fortune cookie to the notion of linguistic texture. After running the program and reading its results, he asked the participants to say what they thought about it. Sandra, a participant in her late twenties who works at an advertising agency as a content editor, said: "At first I thought that it was like some of the earlier works that you showed where there is a corpus like the Bible and then the program selects from it at random. But here there is so much repetition that it starts to lose—it's the same thing over and over again and I'm like, can this repetitious technology have any meaning?" Jason responded: "I think there is very little value in a generator not repeating itself, like having new things always come out. The question is, what type of value, regularity and variation makes a more interesting texture of language? If someone only gets a fortune cookie kind of text all the time, then there is very little value." Jason explains that repetition provides the basis for creating a linguistic texture, as opposed to a single fortune cookie–like decontextualized textual event (e.g., a word or a sentence) that is unrelated to any of the other events that surround it and consequently from which no linguistic texture can be gleaned (cf. Funkhouser 2009:70).[10]

Whereas the poets I worked with view repetition as a key aesthetic dimension of poetry generators' output texts, repetition is not a desirable dimension of such generators' code. Thus, in the workshop's third session, Jason asked Tim, a participant in his late twenties who had just presented the output text of a program he had written based on a prompt given in the previous session, why he had presented the output text but not the code. The following exchange then ensued:

TIM: Well, the code is kind of ugly.
JASON: Let's take a look at it. I would like us to answer the question of what makes code ugly. Why would it be ugly? The texts and the images you presented us were not ugly. I thought they were nice texts. I think they were beautiful images. So why would we refer to the code that produced these as ugly?
TIM: It's just not how you are supposed to write code. I wouldn't want anyone to know I coded like this. Computer science programmers are dissuaded from repetitioning code. If you can't make modules—you really want to write something once and then use it. I would call a code ugly when you are writing the same thing over and over again.

In the discussion that followed, Sid, a graduate student in computer sciences in his early thirties, commented:

Mathematicians and computer scientists will have strong aesthetic values around code and structure, and also teach and are taught these values. The kind of ways we have to talk about the aesthetics of images are not related to the aesthetics of code necessarily. I think the aesthetics of code and of math are kind of about the ways in which this technical system is read. . . . Like a computer science or a mathematician wonder about concision, about following certain conventions, and also removing signs of authorship or subjectivity, which is kind of the inverse of what we are trying to do here with poetry.

Tim's explanation and Sid's elaboration make it clear that a poetry generator's code and its output texts are subjected to entirely different aesthetic regimes because each serves a very different purpose and is considered to be a very different kind of text. Whereas code functions as a means of defining or creating a text distribution, code's output texts function as realizations or samples of this text distribution. Code is assessed in view of the conventions that underlie computer-programming writing practices, where concision and readability are highly prized and repetition of instructions is discouraged (Turkle and Papert 1990:137). In contrast, in the kind of computer-generated poetry that was discussed in the workshop, code's output texts are assessed in terms of the linguistic texture that they exemplify, that is, in terms of a very specific kind of poetics based on repetition and redundancy.

Throughout the workshop, the facilitators argued that because the purpose of writing computer-generated poetry is to create text distributions, the poet should strive to edit and refine only a generator's code in light of its resulting output texts rather than the output texts themselves, since altering or editing the texts would be akin to editing or altering the sampled responses given by the people of Lichtenstein, to use Jason's metaphor. In the writing process, the output texts are approached as indications of the linguistic texture the poet created by means of the code at a specific moment in the writing process, and in light of which the poet might decide to edit or alter the code. This approach exists alongside others.[11] Thus, during our conversation Jason discussed one of the computer-generated poems he had published in a book that showed both the poem's code and its output text. He commented on the orientation that informed the creation of this poem by contrasting it with another possible orientation:

> One approach would be to say, my computer program, the system that I developed, that I use, is like a tool for imagining new things that I can then use in my own writing. Jim Carpenter, in Philadelphia, built a system called Electronic Text Composition and he refers to it as a prosthetic imagination.[12] So it's like your spell checker on Word processor or something. It helps him to imagine. But for me, this [i.e., his own] output text is unedited. It's exactly what comes out of this computer program. It's not that I didn't do any revision or didn't make any editorial effort. I did a tremendous amount of work. I spent a year and a half revising and trying out different versions of this program. But the editing was of the program, not of the final output text.

Whereas an approach such as Carpenter's focuses on developing "systems that tackle the problem of supporting humans in the production of creative linguistic content" by providing them inspiring material that they can then use when they want to create new poetic works (Gatti, Özbal, Guerini, Stock, and Strapparava 2019:242), Jason approaches the output texts produced by any of his poetry generators as realizations or samples of the specific text-distributional essence that is unique to each generator, in light of which he might decide to edit a generator's code in order to modify this essence.

Shifts in the Values of Words, Phrases, and Sentences

The aesthetics of linguistic texture or textual patterns that informs the kind of computer-generated poetry on which the workshop focused produces a shift in the value of the individual word compared with its value in nondigital poetry. This shift has two manifestations. First, words begin to be viewed as raw material for the computer to process. For example, after Jason replaced some of the words that appear in *Stochastic Texts*' code with words taken from Melville's *Moby-Dick*, Allen, a participant, commented: "You are changing the seeds." Jason responded: "Yes, there is a seed and source as concepts and, with pseudo-random number generation, it is kind of like in botany." Individual words thus become "seeds," a term used in specific computational processes such as the genetic-algorithms framework I discussed in chapters 1 and 2. The use of the notion of seeds denotes that in computer-generated poetry, words often come to function and to be viewed as raw material or "data" for the computer to process, based on the specific rules that the poet crafts by means of the code he or she writes. Similarly, consider the vignette with which I opened this book, which took place in the workshop's first session, after Jason read an output text of one of his computer-generated poems. Recall that after Jason described the architecture of the poem's code, Jennifer, a participant in her early thirties who teaches in a master of fine arts program, asked him: "Are you selecting the words?" Here is how this exchange unfolded in detail:

JASON: I'm not doing anything. My computer is doing everything by itself.
JENNIFER: In the Python, I mean.
JASON: Yeah, I just showed you the program. So that's the whole program. It's only sixty lines long.
JENNIFER: But are you choosing the verbage [sic] for—
JASON: The verbage! That's a very poetic way to put it! So I wrote the text because I wrote the program, including all the words [that the program uses].

This exchange revolves around a misunderstanding between Jason and Jennifer. After Jason explains the rules that he coded, Jennifer asks if he chose the words that appear in the output text produced on the basis of those rules. Jason, thinking that Jennifer asked whether he decided where exactly each specific word would appear in the specific output text that he read, that is, whether he edited or wrote the output text in addition to the code, replies, "I'm not doing anything. My computer is doing everything by itself." His denial of any intervention in the resulting output text harks back to the idea of not altering the output texts that a poetry generator produces, but rather using them only as indications of what a generator can do at a specific point in time, which can then motivate the poet to make changes in the generator's code. Jennifer then clarifies that her question revolves around the words that appear in the code itself. The term she uses to do so, "verbage," is significant, as can be seen from the fact that Jason was taken aback by it. After the session ended, I asked Jennifer to clarify the term. "I haven't really thought about it when I said it," she said, "but if I had to justify it now, I would say it's probably a combination of 'verbiage' and 'garbage,'" and smiled. This combination betrays a specific orientation to the words that are used as raw material for the computer to process according to the rules the poet codes, and that appear in the output texts produced when the code is executed. Because those words appear over and over in the resulting output texts when the code is executed, they lose the singular character that they typically have in nondigital poetry. They come to be viewed as a kind of data that only become significant when they are aggregated en masse, at which point they are also frequently experienced as redundant, because one encounters them time and again in different permutations in the output texts. Jason's response to the term "verbage" betrays his discomfort with it because it points, perhaps too obviously, at the shift in the value and function of words in many of the classic computer-generated poems that were discussed in the workshop, namely from vehicles of a qualitatively singular meaning to a kind of data whose value resides in their excess and the general information *qua* texture that can be abstracted from their different (and typically permutational) arrangements.

The value of higher-order textual constructions such as phrases and sentences also shifts as a result of the text-distributional aesthetics that informs computer-generated poetry. Given that a typical poetry generator can, in principle, produce an extremely large number of "events," such as sentences, according to the specific distributional logic this generator represents, no single event or sentence is essential in itself. Often, every event or sentence becomes interchangeable with every other event or sentence within the text-distributional universe of that single poetry generator. This fact found its clearest expression when the guest poets who were invited to the workshop presented and read their work. After a short explanation about the computer-generated work they had written, they ran the programs with their laptops. In some of the computer-generated poems, words gradually appeared horizontally on the large television screen, one after the other, at a slow pace, whereas in other computer-generated poems entire lines appeared, one after the other, in a successive fashion, from top to bottom. This successive appearance of text iconized the notion of text generation, although in principle this gradual appearance of text is not necessary in any computational sense. The pace at which each line or word appeared changed from poem to poem according to the definitions the poets embedded in the code. Although all of the presenters tried to adjust the pace of their reading to the pace of the generated text, they were not always successful. Crucially, they were not bothered by this at all. They would frequently skip lines or words or start reading from the middle of a line. Commenting on his reading of *Stochastic Texts* during the workshop, Jason told me: "If you're reading like the first half of *Stochastic Texts* and then it goes off the screen, then you can just read the second half of the next one [line]. Why not? They are both things that the computer could have presented, right? They are both samples of the same text distribution" or linguistic texture that is the aesthetic essence of this poem (see Strickland 2006:6–7).

The advent of electronic computation accelerates and intensifies the process of viewing words, phrases, sentences, and higher-order textual constructions through the prism of the potential for endless generation and therefore "sampling," although many poets use the computer to facilitate the type of work they have already been doing for many years (Funkhouser 2007:78–79), and although many computer-generated poems are based on procedures that had previously been executed manually (68). In our conversation, Jason commented on this point:

> One of the things you could do with computing, and one of the most powerful things about it, is that it is easy for the computer to do something—almost as easy for a computer to do something five hundred times, five thousand times, as [it is to do it] once. And you can also have the computer do something an unbounded number of times out of just "repeat" until there is a power cut or until you interrupt the process or whatever. So, for instance, you can sample from that distribution, as long as you like, even if it's a finite distribution, like those people from Lichtenstein. You can keep sampling from it, you can keep getting those people.... So these abilities to produce texts in an endless stream are very provocative. It's not that any particular thing is new about it. In a way it is just sped up. But when you speed things up many orders of magnitude

and you automate the process and so on, then you arrive at something that is qualitatively different.

Jason's description of the new possibilities that digital computation opens up for the poet brings to mind Brian Eno's aforementioned argument that creating "music that doesn't have a beginning or an end, that is just theoretically infinite," did not become possible for him "until the 90s when [he] started working with computers." Jason suggests that computation has led to a quantitative leap that in turn has led to a qualitative shift in what the poet can do. The discussion thus far suggests that an important dimension of these changes has been the transformation of words, sentences and higher-order textual constructions from quasi-singular qualities to datalike or quantity-related entities.[13]

Such are some of the aesthetic and technical dimensions of computer-generated poetry, into which the workshop facilitators tried to socialize the workshop participants. As I show in the next three chapters, these dimensions provide the backdrop for a number of culturally and historically specific potentialities and tensions that characterize computer-generated poetry as a distinct form of creative practice.

6

"I Randomize, Therefore I Think":
Computational Indeterminacy and the Tensions of American Liberal Subjectivity

Unexpected Cartesian Meditations on Computer Thought

On the second day of the workshop, Gregory, a participant in his mid-thirties who works at a design firm as a programmer, was about to present an output text of a poetry generator that he had created based on a prompt Jason had given the day before. Gregory approached the media unit, hooked up his laptop to it, and read aloud the following output text, which was projected on the big screen:

> I think, therefore I am.
> I am, therefore I access.
> I access, therefore I choose.
> I choose, therefore I compile.
> I compile, therefore I conform to constraints.
> I conform to constraints, therefore I control.
> I control, therefore I copy.
> I copy, therefore I create.
> I create, therefore I delete.
> I delete, therefore I do not care.
> I do not care, therefore I do not fail.
> I do not fail, therefore I execute.
> I execute, therefore I format.
> I format, therefore I halt.
> I halt, therefore I iterate.
> I iterate, therefore I obey.
> I obey, therefore I print.
> I print, therefore I process.
> I process, therefore I process.
> I process, therefore I randomize.
> I randomize, therefore I think.

The following exchange then ensued between Gregory, Jason, and two other workshop participants, Grant and Laura:

GREGORY: When I started thinking about [writing] this poem . . . I started with the simplest template I could think of, "I think, therefore I am." And at first I wanted to add verbs and make many funny rhymes, but then I realized there was something deeply ironic about having a program create a text, "I think, therefore I am." Because does a program think? And, if so, does it "think" think? Does it contribute anything to being?
JASON (ADDRESSING THE PARTICIPANTS): What sense do you get from this text?
GRANT: It's very stepwise. Every sentence makes sense.
JASON: All right, perhaps there is too much sense. What else?
LAURA: I was also thinking whether the program was needed here. There is a series of verbs but they are used in order. So why use the program?
GREGORY: They are not ordered in a predetermined way. The order is random.
LAURA: Oh, the order is random? Ah, OK.
GREGORY: Yes. I started with a series of words that are somehow related to computing, and the way it works—it always starts with "I think, therefore I am." And then it goes through the list [of words]. In every line, a word is chosen randomly [from the list], and the word that ends one line is carried over to the next line. That always happens.

Gregory argues that this poetry generator is ironic in that its topic is whether and to what extent computers can think. Each of this generator's potential output texts begins with Descartes's famous dictum "I think, therefore I am," which has become a cornerstone of rationalism. Against this backdrop, it is crucial that whereas each output text begins with Descartes's dictum and then follows with a succession of lines describing a number of functions that a computer can perform, the specific output text that Gregory chose to read to the participants ends with a kind of conclusion: "I randomize, therefore I think," which seems to equate computer "thought" with computer-based randomization, at least in the context of computer-generated poetry.

This equation finds validation in the two participants' comments on the poem. Grant's reservations about the poem stem from the fact that it seems to be too orderly and each sentence makes perfect sense. Laura elaborates on this point by saying that the verbs are used in what seems to be a logical order, and hence it is unclear why Gregory had to use a computer to write his poem to begin with. Grant's and Laura's reservations turn on the fact that it is impossible to see the computer's contribution to the poem. Gregory seems to know what specific contribution Grant and Laura have in mind. He clarifies that the verbs in each line do not appear in any predetermined order (save for the first line, the last line that always ends with "think," and the first verb in each line, which is always the same as the last verb in the previous line). Rather, the computer chooses the verbs randomly from a list. Knowing that randomization is a core component of this computer-generated poem assuages the participants' concerns. Gregory's poem can thus be interpreted as a computer-generated poem that uses computer-based randomization as part of its very architecture in order to provide a kind of meta-commentary on the central role played by this form of randomization in "computer thought," when such "thought" is used as a resource in writing this type of poetry.

As I argued in the introduction, the different forms of indeterminacy afforded by computing, which are not limited to randomization, are some of the key reasons poets integrate computing into their practice to begin with (Hayles 2009:43; Taylor 2014:170–171; cf. Helmreich 1998b:75; Turkle 1991:235). However, the poets I worked with do not experiment with computational indeterminacy as part of an aesthetics of "nonintention" (Funkhouser 2007:64) or "authorial abnegation" (Iversen 2010:21), as scholars have argued with respect to the use of chance procedures in digital literature and the literary avant-garde in general. As I show in this chapter, rather than replacing "the desire to do something with the desire to see what will happen" when randomness is integrated into their art (Iversen 2010:24), my interlocutors engage with computational indeterminacy to cultivate different forms of authorial intentionality, as well as to pursue different and often conflicting goals.

First, my interlocutors argue that computational indeterminacy opens up a space for them to intentionally problematize and challenge Romantic notions of creativity whose associated tropes of a rarefied talent and an obscure creative process have provided the basis for different forms of exclusion. Second, they also argue that computational indeterminacy provides them a resource for identifying their own creative habits and fixations so that they can refashion their creative intentionality as a self-governed one, regardless of any social cause that such an intentionality might serve. Third, the specific features of computer-based randomization lead some of my interlocutors to attribute to the computer a form of creative agency that is strikingly similar to the same Romantic tropes of creativity that some of them seek to problematize. Finally, their use of computer-based randomization as a key resource results in a highly rarefied aesthetics characterized by abstract formal experimentation that might provide the basis for taste-based forms of exclusion.

These tensions are not the peculiar feature of the cultural order of computer-generated poetry. Rather, they are the result of, and thus point to, computational indeterminacy's affordances in relation to the different and often conflicting cultural currents that inform contemporary American liberal subjectivity in general.[1] These currents include calls to advance social justice and inclusivity; an emphasis on self-determination; a willingness to approach technology as an inscrutable, autonomous, and self-determining form of creative agency; and the celebration of radical aesthetic experimentation as a form of expressive freedom.

Anthropologists pay close attention to the ways in which human practice informs and is informed by culture, which is never a harmoniously organic whole, but rather a set of norms and practices that are often in tension with one another. These cultural tensions and contradictions may produce unintended consequences of which individuals may be unaware. In itself, then, the fact that there are tensions between the different currents that inform computer-generated poetry is not surprising, since cultural contradictions are characteristic of many organized forms of human practice, including American anthropology.[2] What is important is the specific form these contradictions take in specific ethnographic contexts, and the specific consequences they produce.

Hyperintentionality as a Socially Oriented Response to Computational Indeterminacy of Process

I was sitting in a bar in Manhattan's Meatpacking District with Sid, one of the workshop participants. We discussed a specific session that had taken place earlier that day. Our discussion revolved around the implications of the fact that writing code plays an important role in writing computer-generated poetry. Sid commented:

> The fact that code is so important impacts the writing process in a significant way. First, there is no spontaneity. This is not like regular poetry where you can write a poem in a minute on a napkin in a bar and then give it to someone else to read. Code is about figuring out the relations between different things, of what you are trying to do, of what the outcome might be. . . . Remember what Jason said today: "You have to have a sense of what you are trying to do, and also be able to anticipate what the code you are writing will do." It's way more abstract and slow than the regular use of language. I mean, Python [the programming language used in the workshop] is mathematics! These people have to have an understanding of functions. It is way more difficult and less spontaneous than using regular language. Very few people, even supercoders, can code in real time and know what their code will do without testing it with a computer. Also very few people are able to just look at code and know right

away what the output will look like. This lack of spontaneity—have you ever seen a demo party? It's not some wild thing with music, booze, and fun [Sid points at the people around us in the bar]. There is nothing fun about it. It's usually a bunch of men who sit in front of their laptops, order pizza, and code quietly. This is what it means to write computer-generated poetry.... You cannot do it immediately and be finished.

Sid's comments are informed by a specific language ideology or culturally specific ideas about how different forms of language (including programming languages) function. More specifically, they show "the ways conceptions of language and linguistic practice—indeed of communication more broadly—depend on *differentiations*: the differentiations among signs, among people's social positions and historical moments, and among the projects people undertake" (Gal and Irvine 2019:1). In a related fashion, they also provide an interesting twist on what Derrida has called the Western metaphysics of presence. Derrida has argued that writing has occupied a secondary and inferior status with respect to voice in the context of a specific Western metaphysics of presence (1977). Voice has been one of the key metaphors for the spontaneous, effortless, and transparent expression of thought, compared to which writing has represented a more mediated, degraded, and opaque form. However, as Sid's comments demonstrate, "presence" is a relative concept. Sid's understanding of the relation between, on the one hand, writing and reading "regular language" and, on the other, writing and reading code mirrors and is modeled upon the long-held understanding of the relation between voice and writing. Sid conceptualizes writing and reading "regular language" as a much more spontaneous, immediate, and transparent form of expression of thought and intention—akin to voice—than writing code. He argues that computer-generated poetry is based on a fundamental dimension of computational indeterminacy of process, in that programming is a highly abstract form of expression whose meaning and implications are not as immediately available and clear to the poet as the use of "regular language." Consequently, poets must be highly intentional and deliberate when they write computer-generated poetry, whether or not it incorporates chance processes.

In a conversation I had with Jason, he confirmed Sid's assessment of the challenges that result from the fact that code is computer-generated poetry's foundation, even for "supercoders," to use Sid's phrase. Jason said:

> I think very rarely a display of the code [of a computer-generated poem] is the best introduction for any audience, even if it's a computer science or a programming kind of audience. Actually, when I spoke at Google I did show the code first and then the output, but even then there are very few things that

you can see from that. If I showed you that code, if I showed anyone that, even an expert programmer, if I flashed it out on the screen for ten seconds, no one is going to know what it does. You have to study it for a little while. There are a few challenges in figuring it out. They can tell that it is not just opening a file and reading—they can tell it's self-contained. They can tell it's a small program. And they can see what sort of lexical data is in there, what letters are in there, right? And, so, that's my point in showing it. Not that you can totally understand it but that you know something about the bounds or the limits of the program.

One would expect writers of computer-generated poetry to wholeheartedly resent what they consider to be a highly mediated process lacking the celebrated, mythical, and ideologized spontaneity of writing "regular" poetry, in line with the example of "Lisa, eighteen, a first-year Harvard University student in an introductory programming course," who is also "a poet" (Turkle and Papert 1990:133):

> Lisa wants to manipulate computer language the way she works with words as she writes a poem. There, she says, she "feels her way from one word to another," sculpting the whole. When she writes poetry, Lisa experiences language as transparent, she knows where all the elements are at every point in the development of her ideas. She wants her relationship to computer language to be similarly transparent. When she builds large programs she prefers to write her own, smaller, building block procedures even though she could use prepackaged ones from a program library; she resents the opacity of prepackaged programs. Her teachers chide her, insisting that her demand for transparency is making her work more difficult; Lisa perseveres, insisting that this is what it takes for her to feel comfortable with computers. (Turkle and Papert 1990:133)

Indeed, some of my interlocutors occasionally expressed a similar desire to experience code as a kind of immediate and spontaneous form of expression. For example, in my conversation with Sid in the bar, immediately after he argued that even "supercoders" have a hard time knowing what a code can do just by looking at it, he added: "The future utopia will come when people will just be able to hand one another the code of their poetry and people will be able to read and understand it. That will be the equivalent of [nondigital] poetry today. Not exchanging the output texts but the computer programs." Similarly, at the end of the workshop's second day, Jane, a guest lecturer in her sixties who is one of the pioneers of digital literature, said: "It would be nice if everyone could share code in the same way that . . . you can write a poem on a paper [and hand it out]. . . . I was told it would be true in 2000, but I don't see it happening, not even with my grandchildren."

Similar tropes can be found in other communities that are focused on computation-based creative practice. Thus, later in our conversation, Sid dramatized his argument about how computational indeterminacy affects the nature of writing computer-generated poetry by describing the only event he had ever attended in which a coder coded computer-based art in real time and in a way that was diametrically opposed to the highly intentional manner in which poets write computer-generated poetry:

SID: I only saw real-time coding once in music. It was in the Netherlands. A person was coding music in real time, typing code like crazy to generate music on the spur of the moment.
EITAN: What kind of music was it?
SID: It was noise music.
EITAN: Did it sound intentional?
SID: It sounded like noise. [He laughs.] The coder acted as if he knew what he was doing, knew how each thing he was doing would impact the output, but did he actually know? It was noise!

On the one hand, Sid gave this example as an exception that proves the rule that he was trying to convey to me about the fundamental role that computational indeterminacy plays in computer-generated poetry, which requires that poets be highly calculated and intentional in order to achieve meaningful results. In the music event he had attended, the artist seemed to code in real time and in a way that seemed to follow tropes of immediate creative expression, but because the result, by definition, was based on randomness as an aesthetic ideal, as is the case in the genre of noise music, it is highly unlikely that the coder knew (or intended to know) in any precise way what the specific instructions he coded would achieve.[3]

On the other hand, the music event Sid described can be contextualized more generally in relation to "interactive programming [that] allows a programmer to examine an algorithm while it is interpreted, taking on live changes without restarts" (McLean and Wiggins 2012:149). This type of programming "makes a dynamic creative process of test-while-implement possible, rather than the conventional implement-compile-test cycle" (250). Artists and researchers have been trying to develop "new approaches to making computer music and video animation, collectively known as *Live coding*.... The archetypal live coding performance involves programmers writing code on stage, with their screens projected for an audience, while the code is dynamically interpreted to generate music or video" (ibid.; see also Collins, McLean, Rohrhuber, and Ward 2003). These cultural currents suggest that some code-based artistic practices are explicitly informed by Romantic tropes of immediate creative expression as a desired goal.

Overall, however, my interlocutors approached the computational indeterminacy of process, which underlies the practice of writing computer-generated poetry, primarily as an important resource rather than as a frustrating obstacle that they needed to surmount. Thus, in our conversation Jason continued to argue that the space of indeterminacy that exists between writing (or viewing) code and understanding its meaning encourages him to cultivate rather than abnegate his poetic intentionality:

> I am not particularly inspired when I write. Because I consider myself more of an explorer of language and computation, which I would distinguish from— you know, I am not breathed in by the muse or the spirit. I care about . . . what a tiny program might accommodate. My interest is in what language itself is like and what computation is like. I'm just not trying to express myself in any of my work. That doesn't mean that I do not have intention behind it. What I write is very highly intentional, but it's not intending to tell you about my personal history or my emotional state . . . all of these things that a lot of poets do.

Jason embraces the fact that coding is a fundamental dimension of his writing process because it allows him to intentionally explore what he is more interested in exploring, that is, computation itself and language insofar as it can be represented and manipulated by computation. His rejection of the notion of being "inspired" is a direct reference to (and critique of) normative Romantic ideals of poetic inspiration that emerged in Europe in the mid-eighteenth century and that continue to inform perceptions of creativity today. Such ideals construe some people as being able to produce works of art effortlessly and spontaneously by functioning as the expressive vessel of their inner creative nature, of some external agency, or of some combination thereof (Abrams 1971:184–225). Sid's image of a poet who can instantly write a poem "on a napkin" in a bar also references these Romantic ideologies, whose architects saw poetry as the most "natural expression of feeling" (Taylor 1989:377).[4]

The poets I worked with argued more generally that the computational indeterminacy of process that underlies computer-generated poetry and that requires them to be highly intentional when writing it provides them an opportunity to think critically about and problematize these and similar notions of creativity as a rarefied property that is unevenly distributed among people, and to imagine and promote a more inclusive art world, in particular, and society, in general. Some of them informed their thinking about these and similar issues by contemporary American-based critical scholarship on algorithms, digital technologies, and digital literature, as well as by cultures of free software and hacking outside the domain of art (cf. Coleman 2012; Coleman and Golub 2008; Kelty 2008). Thus, during our conversation Jason suddenly asked me: "Are you familiar with Marjorie Perloff's *Unoriginal Genius* [2012]?" He then explained:

> The idea of creativity has become a kind of bad word and some of it is in reaction to the creative writing programs.... I like to talk about creative computing [rather than creativity] because it is a very good umbrella term that is inclusive of many different practices, not only what is called fine art, not only what's called literary writing but also popular work in the demo scene, in basic games programming, and so forth.... The problem with the notion of creativity is that it can restrict the number of people who can perform this.

Jason draws from contemporary liberal scholarship on digital literature to argue that the craft of writing "digital poetry" provides an inclusive alternative to the exclusive norms and practices that inform many art worlds. He prefers the term "creative computing" to "creativity" because it allows him to designate as "creative" highly different practices and practitioners insofar as they have to do with the novel use of computing (cf. Hope and Ryan 2014:13–15).

On the workshop's fourth day, Sarah, an established poet in her midthirties who was one of the guest lecturers, presented a related explanation for why she values the fact that writing computer-generated poetry is not an effortless and spontaneous process, but rather the design of highly abstract instructions, which involves delay and requires the poet to be highly intentional. Her explanation revolved around Alison Knowles and James Tenney's 1967 classic computer-generated poem *A House of Dust*, with which the workshop participants had experimented on the workshop's first day:

> What I like about this code [of *A House of Dust*] is that it's simple but also very powerful. If you put things [i.e., words] into it, it all comes down to rhetorical choice. And that's something that interests me too, especially with programming that has to do with language—how much it foregrounds, or can foreground, rhetorical choice. The importance of—what it is exactly that you are telling the program to do, what it is you are telling it to repeat. You know, the original code for *A House of Dust* repeats "Negro[e]s," it repeats "American Indians."... So what is it that you are going to repeat? If you think of these digital poems as megaphones, what are you going to amplify? What is your intention? So when I teach undergrad and graduate students, this is something they haven't focused on because they are MFA students. We often talk about process in a very abstract way. We don't always foreground it. So that's something that I like about programming, that it very much foregrounds process ... [and] choice in a way that is often not apparent and talked about apropos writing in general, and in creative writing in particular.

FIGURE 6.1. Some of the books that were available to workshop participants. Note "Algorithms of Oppression" (lying horizontally on top of the books) and "Automating Inequality" (first book on the right).

Sarah informs her thinking with contemporary liberal calls to advance social justice in the United States and with contemporary critical studies of digital algorithms as an infrastructure of social injustice. Elsewhere in her presentation, she mentioned "this book, *Algorithms of Oppression*, by Safiya Noble [2018], where she explores how things that seem to be neutral and unbiased in programming are not." I later encountered this book in the small library that was at the participants' disposal in the workshop's main space, alongside other books that focus on the critical study of digital algorithms, such as *Automating Inequality* (Eubanks 2018), and books that focus on the more technical aspects of computing (see fig. 6.1).

Sarah argues that the reason many of her students do not give much thought to intention and process is that their creative-writing practices are informed by their training as MFA students. In this context of predominantly nondigital creative writing, normative Romantic ideals of creativity as the quasi-spontaneous expression of the artist's self thrive alongside craft-oriented approaches (McGurl 2009; Wilf 2011; Wilf 2013c). They can lead to and justify the creation of artworks that might unintentionally contribute to and reproduce social injustice. In contrast, computer-generated poetry provides a context in which poets can cultivate their awareness of these issues because they experience writing code as an activity that requires them to be highly intentional and deliberate. Sarah demonstrates her point by describing her own awareness of the fact that some of the words that appear in the original code of *A House of Dust* might be offensive if she were to repeat them today when writing a computer-generated poem based on that code.

Hyperintentionality as Self-Determination by Means of Computational Indeterminacy of Outcomes

Poetry generators that incorporate randomization involve computational indeterminacy of outcomes. Some of my interlocutors approached this as a resource for intentionally refashioning their own creative intentionality as a self-governed one. This approach came most clearly to the fore when they used their past oeuvre of nondigital poems as raw data for their code to process.

For example, on the workshop's fifth day, Trish, a woman in her mid-twenties who was one of the workshop participants and who works in an apparel start-up, read an output text of a poetry generator that she had created, which uses some of her past nondigital poetic work as raw material. After she finished reading the output text, one of the other participants asked her, "What does it feel [like] to see your work like this?" "It's weird, uncanny," Trish replied; "it is rereading ourselves to ourselves. I think I can see some of my fixations there, like with grammar." Half an hour later Trish read the output text of another poetry generator that she had created, one that randomly shuffles material from her personal diary and places it in specific sentence structures. After running the program and reading the output text, she said, "My code is showing myself in a mirror, and because it's reconfigured you can see more clearly your constraints, tendencies." Trish argues that writing this form of computer-generated poetry can provide her with key insights on her poetic self because it breaks the organic—and, to the poet, taken-for-granted—nature of her single nondigital poem, for example, which normally prevents her from noticing her poetic "fixations."

Sarah, the guest lecturer, described these insights in greater detail when she showed the participants a code that uses as raw data many of the nondigital poems she had written in the past, breaks these poems down into individual words or phrases, and then recombines them in mostly aleatory ways. Sarah zooms in on the different combinations of words and assesses them according to their novelty and poetic value. "I let it spit out reels and reels of this and sort of recycle or upcycle parts of my work," she said. After a short discussion, she added, "So this [code] is very simple but also powerful in the way it allows me to interact with my work. That would be difficult for me to do otherwise, because when I read a [nondigital] poem that I've written, I know what to expect, I know what comes next. I read what I know is there, so it's hard for me to isolate single lines, sounds, and see them differently." Sarah, too, argues that the poet is so familiar with her nondigital poems that they become transparent in terms of their construction, imagery, and composition, as far as she is concerned.

In contrast to Trish, Sarah argued that instructing the computer to break such poems down into their constitutive units and then randomly reshuffle them allows her not only to become aware of some of her poetic tendencies and fixations, but also to reconfigure them. Thus, during a short break on the workshop's fourth day, I asked Sarah if using her own past nondigital poetic oeuvre as raw material in her digital poetry has helped her to discover something about her poetic style. "Oh yes, it has," she said:

> There are certainly subjects or a variety of words that I gravitate toward, or kinds of language. What's surprising to me is that there are moments when I am using concrete, specific images, but elsewhere there would be a really large abstract thing. And I'm like: "Oh, here is where it doesn't matter how much I push toward specificity and a visual image, I have moments where I leap into complete abstraction." Knowing this about my work, now, as I run this [and get] nothing, feeling like a very vague container, never telling what is inside of that container, I can ask: "Do I want to be a container of possibilities, or do I want to be more suggestive and concrete?" Or, another thing: I am not especially inventive with how I use words. I always arrange the objects, verbs, and subjects in a specific order in the line and the sentence. So that sort of thing I'll try to change. . . . It's a very reflexive practice.

The fact that poets who engage in this poetic practice experience it as a form of critical self-reflection points to the increasingly wide reach of a database-focused logic that represents "a new way to structure our experience of ourselves and of the world" (Manovich 2001:219; see also Hayles 1999). This practice resonates with the growing importance of the datafied self, that is, the increasingly widespread idea that the self is a bundle of preferences and predilections with a distributional logic that can become an object of awareness and change by means of the collection and analysis of enough "data." It can be productively compared with practices of self-fashioning and self-improvement by means of self-tracking technologies that use the body as a source of raw data that can be statistically processed to reveal the embodied self's tendencies and limitations in an attempt to reconfigure and overcome them. Wearable self-tracking technologies have been analyzed as a kind of neoliberal technology of the datafied self, in that "to self-track is to heavily value one's choices and the need to be responsible for them while, at the same time, relieving oneself of responsibility by delegating it to external technology" (Schüll 2016:328). Although some of my interlocutors indeed use the computer not only to track and trace their poetic habits but also as a mechanism to generate new poetic content that can extract them from those habits, others, like Sarah, limit the role of the computer to tracking their poetic habits so that they themselves can overcome them as a form of creative self-determination. Their stance aligns with a neoliberal agency that "requires a reflexive stance in which people are subjects for themselves—a collection of processes to be managed . . . and consciously steered" (Gershon 2011:539). Most important, this form of "break with habit" by means of computer-based chance processes does not coincide with a "break with ego" or "with self-indulgence," which scholars have associated with the use of computer-based chance processes in art (Perloff 1991:150). Rather, it is a new form of ego- or self-cultivation in and of itself, which exists alongside the cultivation of one's authorial agency to advance social justice and inclusion.

Computational Indeterminacy of Both Process and Outcomes as Romantic Creativity

The vignette with which I opened this chapter demonstrates how computer-based randomization, which, in addition to other features, is "almost always found in digital poetry" (Funkhouser 2007:18; see also 79, 83), provided some of the workshop participants one of the most easily graspable justifications for using digital computation in their creative practice, as well as the basis for their perception that the computer can function as a creative agent alongside the poet. The way in which the exchange described in that vignette continued to unfold provides further evidence for the importance of computer-based randomization for the workshop participants.

After a discussion of Gregory's poem, Jason said: "Instead of randomizing [by means of the computer's random function], we can sort the list of verbs [in Gregory's poem] alphabetically to create the same effect." He instructed Gregory how to modify his program to achieve this effect and then said, "Now we have the verbs alphabetized. . . . So now we have a deterministic way of doing it. Let's hear it. Can you read it?" Gregory ran the modified program and then read the output text of the poem in which the verbs that he had assembled appeared in alphabetical order. Jason said: "So, as you can see, you could replace 'randomize' with 'alphabetize.' It's an interesting text. A lot of it seems to make a lot of sense in this order. . . . And I would submit that alphabetization, if you take a bunch of words, a bunch of concepts that are lexically representative, and you alphabetize them, that's just another form of randomness. It's just historically the accident of how they happen to be in our language." At this point Adrian, a participant in his late thirties, objected, "But that's just a form of randomness that is based on how things are!" Adrian's remark suggests that listing the verbs alphabetically would not satisfy him because it would represent a form of randomness that already exists in the world, that is, prior to the writing of the program and to its execution by the computer. He feels that the fact that the computer can randomize a list of words each time anew represents a more radical kind of "newness," and hence that a poetry generator incorporating computer-based randomization produces output texts or stretches thereof that are more radically or genuinely new.[5] Adrian's stance is noteworthy in light of the fact that, as I described in the previous chapter, much of the computer-generated poetry that was explored in the workshop is informed by a text-distributional aesthetics. As Jason emphasized to the participants time and again, the core essence of such an aesthetics is a general linguistic texture or pattern, whether such a texture or pattern is the result of code that incorporates randomness or that produces a deterministic sequence of syllables or words. The output texts that are produced when the computer program that the poet writes is executed should be approached as samples of, or as a window into, the specific text distribution that the poet defined by means of the code he or she wrote. Since the results obtained by using the computer's randomizing function would not be essentially different from the results obtained by using alphabetization as a source of randomness, in that both results would be samples of the same text distribution that Gregory defined by means of his code, Adrian's resistance to the latter method and his strong preference for the former method represent the fetishization of technological means over ends, more generally, and a kind of reification of the computer's creative agency, more specifically.

Scholars have documented similar reactions to computer-based randomization, which turn on its inscrutable and enchanting agentive properties, in different ethnographic contexts such as the design and use of digital gambling machines. Digital gambling machines are endowed with computer chips that are "programmed with mathematical algorithms that execute a game's particular scoring scheme and predetermined hold percentage . . . , working in concert with a random number generator (RNG) to generate its outcomes. Even as a gambling device sits idle, its RNG cycles through possible combinations of reel symbols or cards at approximately one thousand per second. The device is in perpetual motion, indifferent to the presence or absence of players" (Schüll 2012:82). The RNG's quasi-autonomous and inscrutable agency has made digital gambling machines mysterious and godlike as far as both gamblers and operators are concerned (see also Suchman 2007:42; Turkle 1991):

> "One of the great questions in philosophy," begins an article in the gambling magazine *Casino Player*, "is how the body of man, which is mechanical and concrete, can contain the element of mind, which is ephemeral. This has come down to the simple statement that we have a 'ghost' or 'god' in the machine. So too with today's slots." Evoking the ephemeral, ghostlike will of the random number generator, some in the [gambling] industry call this component the Really New God. "The RNG runs on a computer chip, but people act like it's casting a spell," a designer told me. (Schüll 2012:84)

Computer-generated poetry too provides fertile ground for the reification of the computer as embodying an independent, inscrutable, and spontaneous creative agency because of the saliency of computational indeterminacy of both process and outcomes when computer-based randomization is used as a key resource in it.[6] This reified creative agency resonates with the same Romantic and pre-Romantic tropes of creativity as a mysterious, inscrutable, and spontaneous force, which some of my interlocutors seek to problematize. If pre-Romantic theories of poetic inspiration stipulated that an artwork is the creation of a mysteriously divine or supernatural external agency such as the Muses or the gods, for which the artist functions as a vessel (Murray 1989), Romantic theories reproduced the notion of an independent and inscrutable creative agency; but they identified this agency with the artist's inner nature or voice, which expresses itself spontaneously. In this Romantic cultural model, "an inspired poem or painting is sudden, effortless, and complete, not because it is a gift from without, but because it grows of itself, within a region of the mind which is inaccessible either to awareness or control" (Abrams 1971:192). Thus in 1803 William Blake, a key Romantic figure, provided the following description of the origins of his epic poem *Milton*: "I have written this poem from immediate Dictation, twelve or sometimes twenty or thirty lines at a time, without Premeditation and even against my Will; the Time it has taken in writing was thus render'd Non Existent, and an immense Poem Exists which seems to be the Labour of a long Life, all produc'd without Labour or Study" (quoted in Abrams 1971:215). Blake suggests that he created the poem without any premeditation or editorial considerations; that it sprang into being spontaneously and without his intention, and even against his will; that he was possessed by his own creative genius to such a degree that time ceased to exist during the writing process; and, finally, that at the end of this burst of poetic inspiration he was confronted with and surprised by the product of his own genius, as if it were written by someone else.[7]

Some of my interlocutors reproduce these Romantic notions in how they relate to, perceive, and describe the computer's creative agency because of its inscrutable ability to seemingly instantaneously produce new textual constructions, while they reject them as outdated and potentially exclusionary when they focus on themselves as highly intentional poets. This tension is also partly informed by the contradictions of liberal subjectivity, whose fundamental dimensions include not only the Cartesian emphasis on rational deliberation but also normative Romantic ideals of expressive individualism that celebrate emergent, spontaneous, and nondeliberative creativity (Coleman and Golub 2008:267). The conclusion of Gregory's poem—"I randomize, therefore I think"—can thus be viewed as a poetic dramatization of the tension between these two strands of liberal subjectivity and its projection onto the computer.

This tension is likely to become more widespread as the computational technologies that artists use become more advanced and more representative of indeterminacy of both process and outcomes. In an interview I conducted with Gregory, he described the computer's creative agency as follows:

> There is always a distance between what I am doing and the output. The output always exceeds my intent. That distance creates unexpected results. For example, I work a lot with sliders and computer interfaces . . . I say, these are my variables, my configuration, and what happens if I put this slider all the way to the left and this slider all the way to the right? How does that affect the results? How might that inspire change in the work itself? . . . And then something visually interesting comes up that results in something unexpected. And when I use machine learning, that's even more complicated: it allows me to take another step back. I am building the system that builds the system that creates the output. I am the author of the system, but the system is the author of the final output. I am authoring the program to build the output. Now, in terms of the creative intent, I am the author but, again, even if we look at what we did in the workshop, there is the random function, right? I set the boundaries but the system has some leeway, in a way, to make a choice, and I don't know what choice it will make. There is no intentionality, of course, but it [the computer] is doing something independently.

Gregory highlights time and again the computer's ability to surprise him, as when he emphasizes the "distance" between his expectations and the "unexpected" outcomes that always "exceed" his intentions (McLean and Wiggins 2012:240, 247–248). Although he is careful to stress that he is the "author of the system," he construes the computer as an agent capable of "independently" making "choices" that cannot be completely reduced to or determined by its broader context or structure, including the programmer's intentions. He discusses this indeterminacy of outcomes not only in relation to computer-generated poetry that integrates computer-based or user-based randomization, but also in relation to computer-based art that relies on machine learning, a computational architecture that represents a higher degree of opacity and indeterminacy of process (du Sautoy 2019). His comments suggest that as more artists choose to use more advanced technologies in their creative practice, the result is likely to intensify the enchanting properties of "computer agency," in line with the same Romantic notions of creativity as an autonomous, inscrutable, and spontaneous force that my interlocutors seek to challenge.

An Avant-Garde Aesthetics of Potential Exclusion

I would like to return to the output text of Shiv's poetry generator, with which I opened the previous chapter:

>   ilehhiloilehhiloilehhiloohoelodeohoelode
>   odillodeodillodeodillodeeleidehieleidehi
>   odillodeodillodeodillodeeleidehieleidehi
>   odillodeododdehiehoddodeeleidehoidiohilo

I argued that Jason's joking suggestion to let Google Translate determine the language in which the output text is written and that the correct answer is Python 3 points to the causal and hierarchical relation that exists between a generator's code (written in some programming language such as Python 3) and its output texts, as well as to the space of indeterminacy that separates the two (since neither Google Translate nor anyone else can identify the programming language in which the generator was written merely on the basis of the properties of its output texts).

On another, more important level, the absurdity of Jason's suggestion, which delighted everyone, stems from the fact that the primary value of the output texts of many poetry generators is not referential, and that many such texts cannot even be traced back to any known natural language to begin with. Although Shiv's poem is an extreme example of the disavowal of the referential function, this dimension has been salient throughout the history of computer-generated poetry, and it can be detected in works as early as the very first computer-generated poems. Recall again the output text of Lutz's *Stochastic Texts*, which I discussed in the previous chapter (see p. 147). Although each individual sentence in it makes some referential sense, the output text as a whole does not. The aesthetic disavowal of referential value is the basis for the two participants' reservations about Gregory's poem, as described in the vignette with which I opened this chapter. Recall that the two participants target the fact that "every sentence [in Gregory's poem] makes sense," as Grant puts it, and that those sentences are arranged in what seems to be a perfectly logical order, as Laura suggests. Such reservations point to the fact that Gregory's computer-generated poem is an outlier in terms of its referential coherence, compared with most computer-generated poems. Overall, "randomization, patterning, and repetition of words, along with discursive leaps and quirky, unusual semantic connections, are almost always found in digital poetry" (Funkhouser 2007:18).

The aesthetics that informs the disavowal of referential value, and often of any semantic value, is multiply determined and historically specific. Early members of the Oulipo—the literary group that some of the workshop facilitators presented as their direct precursors—appropriated structuralism and Lacanian psychoanalysis, the latter being "indifferent to deep meanings, concerned more with a latent organization of the manifest than a latent meaning beneath it" (Duncan 2019:94). In this framework, language becomes important insofar as it "may be treated as an object in itself, considered in its materiality, and thus freed from its subservience to its significatory obligation" (Bénabou 1998:41). In an essay written in 1973, titled "A Brief History of the Oulipo," Jean Lescure explicitly references Claude Lévi-Strauss's notion of "the science of the concrete" to argue that "language is a concrete object" (Lescure 1998:35) and that "[literary language] doesn't manipulate notions, as people still believe; it handles verbal objects and maybe even, in the case of poetry . . . , sonorous objects. . . . Unusual designations [i.e., combinations of words] point to the sign rather than to the signified" (36). It is not surprising that these principles resonate with what Roman Jakobson called "the poetic function" (1960), which refers to language that reflexively points to its material properties, for Jakobson and Lévi-Strauss influenced one another during the same period in which the Oulipo was institutionalized (Geoghegan 2011). These principles express an aesthetic fascination with the sonorous and visual patterns and textures that can be created by treating language as a concrete object apart from and, indeed, often at the expense of its referential function. It accounts for the "venerable history" of the practice of "homophonic translation" among members of the Oulipo, that is, of "translating for sound rather than sense" (Duncan 2019:22), which often required "a generous reader to force [the resulting] lines to function as a fluent narrative . . . [rather than] a confrontational exercise in Dadaist randomness" (62).[8]

As the workshop progressed, it became clear that some of the participants were unwilling or unable to function as "generous readers" of the often nonsensical output texts generated by their and others' poetry generators. Sitting with Leandra, a workshop participant in her mid-twenties, during a break on the workshop's fifth day, I asked her what she thought about the kind of computer-generated poetry that the facilitators had presented in the workshop and that the participants had experimented with. Leandra thought for a few moments before saying: "I find it sort of unfascinating, especially when it starts to sound like just a bunch of sounds. Sometimes the meaning is stripped if there's not a lot of intention put into it. . . . I think it's very easy to write a program that is just going to jumble through your words and make random sounds. It is harder to still make random combinations of words but also have them be surprisingly evocative of a narrative or a feeling." Leandra describes the kind of poetry that was explored in the workshop in terms of random and often nonsensical combinations of words and sounds. Although these combinations can at first glance pleasantly surprise the reader because of their novelty, they can quickly frustrate her once she becomes accustomed to their surface-level novelty and realizes that they lack "a narrative or feeling," or what she elsewhere in the interview called "substance."

One way in which the workshop facilitators responded to the discomfort experienced by some of the participants with respect to the textual outputs of the computer-generated poetry discussed in the workshop was to highlight the aesthetic value that linguistic textures and patterns might have, even though, and sometimes precisely because, they do not display a clear referential meaning or a narrative development, as I discussed in chapter 5. Recall that Jason explained to me that by virtue of creating such textures, computer-generated poetry can clarify and convey specific types of phenomena and experiences such as "looking at a river," which, even though it is "neither surprising nor suspenseful," that is, "less like a narrative event," is still meaningful. The experience of observing a flowing river can be aesthetically provocative precisely because of the kind of repetition or texture that one can observe in this flow.

However, during the workshop it appeared that some of the participants found it difficult to accept the notion of textual patterns as a sufficient aesthetic rationale for computer-generated poetry, and that even when some of them were willing to accept the idea that creating textual patterns might be a valid rationale for computer-generated poetry, they separated this idea from any notion of meaning altogether. Thus, when I asked Gregory how he thinks about meaning in computer-generated poetry, he replied: "Well, right now it may be very small, buried. We haven't really talked about meaning at all. We generate patterns with words, we don't generate meaning with words. Meaning can emerge accidentally, or perhaps somewhat intentionally, but it's very rare." Gregory argues that the workshop focused on generating "patterns with words" rather than meaning, thus referencing the text-distributional aesthetics that the facilitators emphasized time and again, which is based on repetition with some variation. When meaning emerges, it does so "accidentally" and "rarely." Computer-based "stochastic" or "pure contingency" thus becomes one of the most visible distinguishing features of the kind of computer-generated poetry that was the focus of the workshop, alongside a high degree of "semiotic contingency," or the potential unpredictability of meaning and interpretation of any action (Malaby 2007)—an unpredictability that in this art form can border on complete nonsense, as Shiv's poem demonstrates.[9]

When Jason asked the participants to think of questions they might have in relation to the contents of the workshop and then write them on the whiteboard for everyone to see, a few of the questions critically engaged with the idea of creating textual patterns as a sufficient artistic goal. For example, one participant wrote: "Is it possible for computation to be unpoetic?" This question suggests that computation can never be "unpoetic" because it can so easily be used to produce textual patterns. Furthermore, the fact that almost any kind of text can be considered a textual pattern raises the question of aesthetic quality and discernment, as some of the participants' other questions suggested, such as: "What is a 'good' poem?" "What is poetry?" "What is the text?"[10] Jason responded to these last three questions: "These are the sorts of questions you might get asked at the end of your PhD defense, just because the examiners want to have some fun and see what you will say. . . . To start approaching these questions you should break them down and figure out why you want to know these things—what impact on you will it have? Are you asking what a 'good' poem is so you can conform to other people's notions of goodness? Or do you need to develop your own idea?" Jason avoids giving any definitive or prescriptive answer to the participants' questions. He responds by posing his own questions, which shift the focus to the participants and to their motivations for wanting to know whether computation can ever be unpoetic, what is a good poem, what is poetry, and what is a text. He encourages them to figure out why they "want to know these things" and whether what is at stake is their desire to "conform to other people's notions" of aesthetic value. He ultimately suggests that they need to formulate their own aesthetic notions.

Jason's stance is informed by another current of American liberal subjectivity, one that takes the modern avant-garde and its abstract formal experimentations as iconic forms of expressive freedom. Following the World War II and the onset of the Cold War, the modern avant-garde (especially its abstract expressionist instantiations) came to be accepted as a constructive force against both fascism and communism. Its transformation "into aggressive liberal ideology" was made possible, in part, "because the values expressed through the works were especially important to liberals during the Cold War period. Among those values were individualism and the willingness to take risks, central elements in the creative system of the avant-garde and warrants of its complete freedom of expression" (Guilbaut 1983:200). Jason's advice to the participants not to conform to existing aesthetic criteria but rather to establish and follow their own criteria in relation to the highly rarefied abstract formal experimentations that characterize computer-generated poetry is a textbook example of these liberal values (Wilf 2012:32).

However, these liberal values of complete self-determination and self-expression, made iconic by an aesthetic break from the conventional world and the creation of new, abstract formal-experimental worlds, might undermine other liberal values. The different opinions that found expression in the workshop suggest that computer-generated poetry might reproduce an aesthetics of exclusion, despite some of its proponents' avowed commitment to promoting new forms of inclusion in art. The ability to enjoy a nonsensical output text such as the one produced by Shiv's poetry generator or even Lutz's *Stochastic Texts* is the ability to symbolically display what Pierre Bourdieu, in an essay titled "The Aristocracy of Culture," called "distance from necessity" (1980:251). Bourdieu argued that some differences in people's aesthetic predispositions and aptitudes in art are the concealed result of the unequal distribution of economic capital, which was responsible for their different modes of socialization that make it easier for some and harder for others to symbolically display distance from economic necessity by adopting a Kantian aesthetics of "detachment, disinterestedness, [and] indifference" (239). This aesthetics prioritizes formal experimentation and abstraction over "vulgar" works of art that display a direct connection to the immediately available world in the form of clear referential meaning, a logically coherent narrative, figurative representation, and so forth (ibid.; Wilf 2014b:403). To be sure, Bourdieu's theory has been criticized for implying a deterministic connection between socioeconomic background and taste. At the same time, his arguments about the essentially luxurious nature of the kind of abstract formal experimentations, which characterizes important strands of modern high art and that also finds a clear expression in many computer-generated poems, coupled with the fact that the textual results of such poems divided the workshop participants, point to the potentially exclusionary aesthetics that writers of computer-generated poetry who inform their practice by liberal values of complete aesthetic freedom, as made iconic by radical formal experimentation aided by computer-based randomization, might reproduce rather than challenge.[11]

I Randomize, Therefore I Think in Multiple Liberal Ways

My interlocutors' experiments with computer-generated poetry suggest that different and often conflicting cultural currents can inform not only different computer-based creative practices, but also a single such creative practice. These conflicting currents, which in the case of computer-generated poetry reflect the potentials and limitations of American liberal subjectivity, mean that it would be unwise to characterize any creative practice that incorporates chance processes as expressing a single aesthetic principle such as "bypassing of intention," "authorial abnegation" (Iversen 2010:21), or "nonintention" (Funkhouser 2007:64). My interlocutors respond to computational indeterminacy by being hyperintentional and socially oriented when they code, using code as a resource for refashioning their creative intentionality as a form of self-determination in and of itself, attributing to the computer a Romantic form of creative agency, and producing a highly rarefied and potentially exclusionary avant-garde aesthetics as a form of expressive freedom.

Gregory's poem, with which I opened this chapter, speaks to this multiply determined and culturally specific relationship between computational indeterminacy and notions of creative intentionality and agency, despite the fact that it is an outlier in terms of the referential coherence characterizing its output texts. Not unlike the Cartesian dictum that inspired it, it thus encapsulates in an inchoate form a number of important implications whose elaboration, however, begs for a cultural rather than a philosophical analysis. Whereas this chapter has analyzed those implications in relation to computer-generated poetry's present cultural context, that is, American liberal subjectivity, chapters 7 and 8 do so in relation to one dimension of this art form's cultural and intellectual history, that is, its relation to the French literary group the Oulipo.

7

Analog Precursors and Their Digital Logical End:
The Oulipo

The Game of Chance

It was an hour into the workshop's first day when Jason turned to the participants and said: "Before the break, a noncomputer writing exercise for you to do. Take a pencil or a pen. At the top half of the paper write some question, any question that comes to mind. Write it immediately, don't think about it." While the participants were writing their questions, Jason continued: "Having done that—you are taking too long!" he scolded them, "I want you to take the top where you wrote the question and then fold that over. So now there is a blank bottom half. You will exchange your paper with someone else. Don't look at what's been written. Just look at the blank spot." The participants did as they were told. "Now write an answer, some answer, but don't think about it. Write whatever answer comes to mind." The participants complied again. Jason continued: "Now hand this paper to someone else. Let's hear our questions and answers." One after the other each participant read the question and the answer written on his or her piece of paper. The question-and-answer pairs included "Is there coffee? Because it's hot"; "How are you? Southern times"; "When am I going to switch teams? Because I feel safe"; "Why random? Yes"; "What's for dinner? Three sacks of salt"; "Why is the weather? A paragon of human machine"; "Where's the defense? Somewhere through these hallways"; "Who are you? Peace"; and "What? Nomad." As they listened to some of the combinations, a few of the participants laughed.

Jason instructed them to play this game two more times. After everyone finished reading their question-and-answer pairs for the third time, he said:

JASON: So, do these get any better? I don't know.... This is a surrealist writing game, a structure that was used by the surrealists. I don't invite people to do this because it produces good writing. I invite people to do it because

I think it invites people to think about process and collaboration. This is collaborative writing, isn't it? We worked together to make all of these texts. And I want to suggest something that might be controversial. I want to suggest that we all read each other's writing very closely.

SANDRA: What do you mean "closely"?

JASON: Well, obviously we didn't read the question that somebody else wrote when we wrote an answer, right? But I think we were all very interested to hear what the question-answer pairs were when we went around and read them. So even though we didn't immediately write an answer that responded to the question, we did listen to what each person had to write on the question side and on the answer side and how they fit together or not.

This game, which Jason chose to introduce at the very beginning of the workshop, immediately after he presented a number of classic computer-generated poems and before the participants had had a chance to experiment with them, encapsulates both some of the key aesthetic dimensions and tensions that informed the workshop, as well as some of the participants' reactions to them. First, the game is based on a procedural constraint that involves a series of clear steps. Second, this constraint integrates randomness, in that the questions and answers are written independently of one another and are then paired randomly. Third, Jason argues that the participants were curious to know what the results of this randomness-based constraint would be. Fourth, he suggests that it is possible to learn to "feed" this constraint raw material that will produce more interesting results. The reason he made the participants play this game three times in a row is that he wanted to show them that they could quickly learn to write questions and answers that would produce more interesting results when those questions and answers are paired together randomly. In a discussion during the workshop's third day, Jorge, a guest lecturer in his mid-thirties who was present when the game was played and who had collaborated with Jason on a number of projects in the past, made a similar suggestion: "I think the exercise we did the other day with Jason, the questions and the answers, was interesting because every time we did it we were more conscious of how it worked. We gradually tried to give answers and questions that were more likely to work with other questions or answers. This is something that we do when we think about combinations and rules in digital literature."

Fifth, some of the participants quickly became ambivalent about the poetic results of the game. This ambivalence found expression in two ways. First, as the game progressed, fewer of the participants laughed at the randomly paired and mostly nonsensical combinations of questions and answers. They seemed to become inoculated fairly quickly against the poetic effects of this randomness-based constraint. Indeed, the question-and-answer pairs that continued to elicit laughter were those few that made sense, even though they were the product of a random process. Such pairs included "When will I die? Probably over there"; "Are you following me? Wherever you are"; "What's for dinner? Three sacks of salt"; and "Your place or mine? I was hoping you'd ask that." The participants would not have given these pairs a second thought had they encountered them outside of this writing exercise. Second, as the game progressed, some of the participants began to use the game to ask meta-level questions about its purpose and rationale. Such questions included "Why random?," "What's the point of all of this?," and "What is art?"

The first four points turn on the ways in which computer-generated poetry provides fertile ground for the prioritization of technological and computational-algorithmic means over their poetic ends and for the reproduction of the notion of creativity as an opaque, unpredictable and spontaneous quality. Such prioritization and reproduction are the result of the crucial role played in this form of poetry by self-fashioned constraints that are written in code, the poetic potential that these constraints represent, and the computer-based randomness that they often incorporate into their logic. The last point turns on the poetic results of using the computer as a key creative agent in computer-generated poetry, and on some of the participants' ambivalence toward and subsequent pushback against these results and the aesthetic dimensions that inform them.

The game that Jason instructed the participants to play not only encapsulates some of computer-generated poetry's key aesthetic dimensions and tensions; it also points to their intellectual history. When I asked Jason to clarify why he had asked the participants to play the question-and-answer game, he said: "The question-and-answer exercise is a precursor to the Oulipo's work with computing and combinatorics and process and procedure—things that are applicable to digital literature. The Oulipo was a group of writers and mathematicians and computer scientists. They were really the second wave surrealists. And here's a fun fact for you," Jason said with a smile: "Marcel Duchamp was an official member of Dada, an official member of the surrealists, and an official member of the Oulipo." Jason's response points to a specific intellectual history that I have already alluded to and that has played a sporadic role in the previous two chapters. The Oulipo—an acronym for Ouvroir de Littérature Potentielle, "Workshop of Potential Literature"—is a literary group founded in France in 1960. It was mentioned a number of times during the workshop.[1] In many ways, the computer-generated poetry that was explored in the workshop is the digital logical end of the aesthetic principles heralded by the Oulipo, whose members lamented the fact that in the group's early, formative years they did not have computers at their disposal, and who "were among the first to focus on computers as creative supports" (Saemmer 2010:164).[2]

In this and the next chapter I analyze the extent to which some of the aesthetic dimensions, tensions, and contradictions that structure computer-generated poetry as a cultural practice can already be detected in the Oulipo's aesthetic principles and practices prior to the group's turn to using computers. On the one hand, this analysis points to the implications of using digital computation for poetic purposes, that is, to the difference that a digital difference makes. On the other, it also invites a reevaluation of the Oulipo's aspirations, because it reveals that many of the tensions and contradictions that inform computer-generated poetry already inhered in the Oulipo's orientation to creativity as something that can be simulated by mathematical and algorithmic means, irrespective of the availability of digital computation.

A Digital Difference That Makes a Difference

There are many continuities between the Oulipo and computer-generated poetry. First, one of the Oulipo's key principles is "synthesis," or the creation of self-fashioned formal literary constraints that are productive of a large number of different literary outputs (Motte 1998:1). This principle aligns with the core practice of the kind of computer-generated poetry that was the focus of the workshop, namely, writing code by means of which the poet creates a text distribution that in principle can find expression in many output texts *qua* samples of this distribution.

Second, the Oulipo emphasizes mathematically informed kinds of constraints (Motte 1998:14). Its members are either mathematicians or people who are interested in mathematics as a source of ideas for new constraints. Their alignment with mathematics resonates with the mathematical foundations of code in general and of Python—the programming language with which the workshop participants experimented—in particular. Indeed, during the 1960s—that is, in the formative years of digital poetry in general, and computer-generated poetry in particular—"authors programmed poems using coding that was previously designed for mathematical and scientific calculation" (Funkhouser 2007:26).

Third, historically, one of the reasons members of the Oulipo emphasized the importance of self-fashioned constraints that are mathematically informed and rigorously constructed is that they rejected the Romantic notion of poetic inspiration as the basis for literary creation. They were especially averse to the notion of the isolated genius who can create only when he or she becomes unexpectedly possessed by his or her creative nature. The group's decision to oppose "*ouvroir* [workshop] to *oeuvre*, labor to inspiration, collective effort to individual genius, the artisan to the artist" (Motte 1998:10) mirrors the workshop facilitators' stance that I explored in chapter 5, namely, Jason's insistence that he is not inspired when he writes computer-generated poetry and his argument that computer-generated poetry is an inclusive kind of literary practice, Chris's emphasis on this practice's collaborative nature as well as his invitation for other poets to alter his computer-generated poems, and the very organizational form of the workshop in which I conducted fieldwork—an organizational form that the Oulipo carefully chose in order to highlight the group's emphasis on collaborative, experimental, and conscious (i.e., nonpossessed) literary creation (Arnaud 1998:xiv–xv).

Fourth, key members of the Oulipo argued that "the efficacy of structure—that is, the extent to which it helps a writer—depends primarily on the degree of difficulty imposed by rules that are more or less constraining" (Motte 1998:11). They suggested that there is a direct relation between the difficulty of a constraint and its potential literary value. This argument aligns with Jason's description of his poetic style as one that is based on writing under "very difficult constraints that are on the edge of possibility," as I discussed in chapter 5.

Fifth, when the Oulipo was founded, its members emphasized the importance of creating constraints rather than exploring their literary implications. Raymond Queneau, the group's cofounder, argued that "once a constraint is elaborated, a few texts are provided to illustrate it. The group then turns to other concerns, and the texts thus engendered are disseminated to the public" (Motte 1998:12). Other members maintained that "the ideal constraint gives rise to one text only," that is, "a constraint must 'prove' at least one text" (ibid.). Most extreme were those members who insisted "that the only text of value is the one that formulates the constraint: all texts resulting therefrom . . . must be banished to the limbo of the 'applied Oulipo' " (ibid.). These aesthetic principles point to a clear hierarchy that privileges the self-fashioned constraint over its realization.[3] This hierarchy resonates with the workshop facilitators' emphasis on the greater importance of code as a kind of deep underlying self-fashioned structure by means of which the poet creates a text distribution, compared with the output texts that can be produced by executing the code, which are approached as a surface-level kind of phenomenon whose main function is to reflect or be samples of this text distribution.

These continuities between the Oulipo and computer-generated poetry are intertwined with a number of significant differences. To begin, the reason members of the Oulipo gave for the greater importance of the self-fashioned constraint compared with its realization are different from the reason the practitioners of computer-generated poetry I worked with gave for the greater importance of code compared with its resulting output texts. For members of the Oulipo, producing more realizations of a specific constraint does not align with the experimental nature of the group and with its battle against literary ossification. They argued that whereas a "traditional constraint [such as the sonnet] . . . presupposes multiplication, and even demands it," thereby tending "towards imperialism" (Roubaud 1998:91), an Oulipian constraint "can tend toward multiplicity . . . only in ceasing to be Oulipian" (92). In contrast, the reason the workshop facilitators gave for not executing time and again the code of the kind of poetry generators that were discussed in the workshop is that because of these generators' text-distributional aesthetics, only a limited number (or length) of their output texts is needed in order to get a sense of these generators' aesthetic properties, that is, the linguistic texture they can produce. As the participants learned during the workshop, any additional output texts would not be significantly different from the output texts that were already produced the first time the code was executed.

The explanation that the workshop facilitators gave for why it would make little sense to generate numerous output texts by means of a specific poetry generator problematizes one of the justifications given by members of the Oulipo for their literary practice. Members of the Oulipo celebrated the fact that their self-fashioned constraints can generate a vast number of different literary texts. This justification finds its epitome in a text composed by Queneau, titled *One Hundred Thousand Billion poems*, which, according to members of the Oulipo, crystallizes the Oulipian spirit. It consists of ten sonnets. Each line of each sonnet can be combined with each line from the other nine sonnets, thereby giving rise to 10^{14} different potential sonnets (since a sonnet has fourteen lines). The following description of Queneau's poem is indicative of its status in the Oulipian world:

> Everything had begun with the *Cent Mille Milliards de poèmes*, which Raymond Queneau was in the process of composing. When this composition was finished, the work was hailed by the Oulipians as the first work of potential literature. . . . One sees . . . all that makes for the potentiality of the *Cent Mille Milliards de poèmes*: it is not only the example, the archetype they constitute, but also the ninety-nine trillion nine hundred ninety-nine billion nine hundred ninety-nine thousand nine hundred ninety sonnets that are found, inexpressed but *in potential*, in the ten others. (Bens 1998:65)

This description points to the fetishization of the high number of potential texts and of the Oulipian self-fashioned constraint that can generate them, a fetishization that also finds expression in the decision to represent this high number by means of words rather than numbers—a decision that significantly lengthens this representation.[4] The workshop facilitators' take on the text-distributional logic that such constraints embody and their argument that exploring all their implications would be a futile exercise that would only yield more of the same, expose this fetishization. Queneau's poem may be productive of a very large number of different potential manifestations, but continuing to realize them would yield diminishing returns.

Of course, when Queneau wrote his poem, the use of machines was still an unrealized dream. Queneau argued that the "craftsmanlike" dimension of the Oulipo "is not essential," as it could be performed by "machines," the lack of access to which "is a constant *lamento* during [the Oulipo's] meetings" (Queneau 1998:51). The workshop facilitators' stance is informed by the fact that, as opposed to Queneau, they have at their disposal an agent—the digital computer—that can easily explore and present them with the numerous implications of a given constraint. Had Queneau had at his disposal a machine capable of presenting him with a large number of the output texts that can be produced from the constraint he had devised, perhaps he too would have emphasized less the importance of their immense number.

Crisis and Clinamen

If the workshop facilitators' argument that it would be futile to explore all the potential implications of a given constraint problematizes the profundity of the Oulipo's earliest foundational texts, the facilitators' insistence that they not edit a poetry generator's output texts but only its code is problematized by the Oulipo's most famous texts, written by members of the group's second generation. Those members wrote texts based on constraints whose implications do not clearly derive from them and whose satisfying exploration depends on the poet's creativity rather than on any kind of automatic procedural execution that can be performed by a digital computer. These later texts are based on aesthetic principles that took shape during a crisis among the group's members in the mid-1960s.

The Oulipo's first generation of writers, headed by Queneau and François Le Lionnais, argued that the purpose of the Oulipo is to create mathematically informed constraints and procedures that can be followed mechanically and without any deviation and that can function as the means for mining a trans-individual form of the unconscious that can be found in language as a concrete object. As I argued in chapter 6, they were influenced by Jacques Lacan's ideas about the unconscious as language, as well as wider structuralist ideas about the "death of the author" (Barthes 1978). Their goal in crafting and rigorously following procedures of language manipulation on the syntactic level was to "evacuate the subject from writing—to 'let language itself speak'" (Duncan 2019:24), and thus to create literature that was not beset by what they considered to be false Romantic ideas about the cardinal creative importance of the individual author and his or her repressed and capricious creative unconscious. It was easy for them to envision a future in which machines would be responsible for the task of following and exploring the implications of such formal literary procedures, because they focused on creating strict procedures for language manipulation on the syntactic level rather than on higher levels such as those of plot and narrative (Arnaud 1998:xv; Duncan 2019:108).

In 1966 the group experienced a crisis caused by "frustration at the narrowness of the group's interests" (Duncan 2019:23). This crisis led to the election of new members and to a shift of focus, from creating strict procedures of language manipulation to crafting more open-ended constraints whose purpose was to help individual authors express themselves and their creativity by producing literary texts that displayed plot and narrative. If Queneau's *One Hundred Thousand Billion poems* is the text that epitomizes the first generation's aesthetic principles, Georges Perec's lipogramatic novel *The Disappearance* (1969)—a novel written without the letter "e," the most frequently used letter in French—epitomizes the second generation's aesthetic principles. Although the lipogramatic constraint restricted Perec in many significant ways, one cannot say that his novel is a direct implication of this constraint or that it can be deduced from this constraint in a mechanical way, as opposed to the clear connection that exists between Queneau's poem as a kind of algorithm, and all of its resulting output sonnets. Perec's genius consisted not so much in inventing new constraints (indeed, the lipogram is an ancient constraint that Perec appropriated) as in his ability to write creative literature and poetry based on those constraints.

Inasmuch as the founders of the Oulipo called the kind of literature to which the group is devoted to writing "potential literature" (Queneau 1998), the difference between the Oulipo's first generation and second generation of writers can be conceptualized by distinguishing between different cultural notions of potentiality. Karen-Sue Taussig, Klaus Hoeyer, and Stefan Helmreich (2013) identify three meanings of potentiality that are prevalent in popular, scientific, and anthropological texts: (1) "A hidden force determined to manifest itself—something that with or without intervention has its future built into it"; (2) "Genuine plasticity—the capacity to transmute into something completely different"; and (3) "A latent possibility imagined as open to choice, a quality perceived as available to human modification and direction through which people can work to propel an object or subject to become something other than it is" (S4). The Oulipo's first generation of writers adhered to the first meaning of potentiality. They argued that the different potential texts that can result from the exploration of the constraints they developed inhere in those constraints and can be arrived at by a simple and mechanical deductive process, which renders those texts almost uninteresting. In contrast, the second generation adhered to the third meaning of potentiality: they crafted constraints whose potential literary products inhere in them in a very inchoate way and require the active and creative intervention of the poet in order to be realized.[5]

The desire to experience and exercise their creativity motivated a number of the Oulipo's second generation of writers to reject the idea that the computer might eliminate the craftsmanship needed for exploring the implications of specific constraints (Roubaud 1998:85). In addition, it led them to insist that they should be allowed to deviate in small ways from the constraints that they devised, which already allowed for much more authorial freedom than the rigid procedures devised by the Oulipo's first generation. They appropriated the Lucretian notion of the clinamen to theorize such a deviation and to explain its necessity. Motivated by what one member of the Oulipo, Jacques Roubaud, described as their "disappointment at not being able to use such-and-such a word or image or syntactical construction that strikes one as appropriate but is forbidden," they appropriated the notion of the clinamen as "an intentional violation of constraint for aesthetic purposes" (quoted in Duncan 2019:118). Consider this description of the function of one form of clinamen that Perec devised in writing one of his pieces: "Its role is analogous to that of the joker in certain card games, in that, from the moment it is introduced, a satisfactory result is more easily obtained. . . . Its consequences are considerable: the language of the new form, when compared to the old [i.e., when there is no deviation from the formal constraint], describes a radical swerve toward the normative, the texts engendered are more *readerly*, less *writerly*. . . . [Perec's] clinamen is patently a locus of free will, a reaction against the constraints (self-imposed, in Perec's case) of a rigid symmetry" (Motte 1986:275).

The clinamen's role is thus twofold: it not only creates a space of creativity for the poet by allowing him or her to break free from the strict implications of a constraint, it also makes it possible—by means of this creativity—for the resulting texts to be more aesthetically pleasing and relatable as far as the reader is concerned: "From [Perec's] initial belief in the value of maximal formal rigor as the guarantor of the text, he came to feel that the textual system must be intentionally flawed, the flaw scrupulously cultivated, in turn, as the real locus of poetic creativity" (Motte 1986:276).[6]

The emphasis placed by the Oulipo's second generation of writers on the dual importance of the clinamen casts a critical light on the emphasis placed by most of the workshop facilitators on the importance of not altering or editing a poetry generator's output texts, as opposed to the generator's code, as a kind of self-fashioned constraint. As I discussed in the previous chapter and as I will further explain in the next, one of the most significant reservations that some of the workshop participants had regarding computer-generated poetry resulted from this emphasis and mirrored the discontent felt by the Oulipo's second generation. This reservation turned on the fact that many of the output texts of the poetry generators created by the workshop participants were difficult to make sense of, at best, and completely nonsensical, at worst.

There is one additional unexpected common denominator that unites the Oulipo and the practitioners of computer-generated poetry with whom I worked. The Oulipo replaced their early argument that poetry and literature could be mathematically formulated in the form of carefully calculated constraints whose implications can be automatically deduced from them with the expectation that the group's members would creatively explore the possible implications of their self-fashioned constraints. This shift in emphasis created fertile ground for the reemergence of a Romantic poetic aura centered on the notion of the poet's unique and creative genius, despite the group's avowed resistance to this notion. The Oulipian poet or writer has reemerged as a locus of artistic singularity, originality, and creativity as opaque and mysterious qualities—only now, such singularity, originality, and creativity find expression in his or her ability to create literary texts of worth based on highly demanding constraints.[7] Hence the following definition of the Oulipo, "proposed by the group in its early days," which is supposed to encompass all of its members, more accurately applies to the group's second than to its first generation of writers: "rats who must build the labyrinth from which they propose to escape" (Motte 1998:22). Although the practitioners of computer-generated poetry I worked with "propose to escape" "the labyrinth" that they "build" by means of the digital computer rather than by themselves (for they assign to the computer the task of exploring the space of possibilities opened up by the set of constraints they define by means of code), their reliance on the computer also leads unexpectedly to the reemergence and reproduction of the ideologized notion of creativity as an autonomous, inscrutable, and spontaneous force, despite their avowed resistance to this notion. In their case, as I argued in chapter 6, such reemergence and reproduction take a different and much subtler route, namely that of computer-based randomization that results in the reification of the computer's creative agency.

Chance, Anti-Chance, Digital Chance

The digital reproduction of the notion of an independent, inscrutable, and spontaneous creative agency in the cultural context of computer-generated poetry casts an ironic light on the following prediction by Italo Calvino, one of the Oulipo's second generation of writers: "The aid of a computer, far from *replacing* the creative act of the artist, permits the latter rather to liberate himself from the slavery of a combinatory search, allowing him also the best chance of concentrating on his 'clinamen' which, alone, can make of the text a true work of art" (Calvino 1998:152). Using the computer in a form of poetry that is based on self-fashioned constraints has in fact replaced parts of the artist's creative act. Rather than exercising their own creative agency by creating literary texts that explore the implications of, but that also deviate from, the constraints that they fashion, as Calvino hoped would be the case, some (though, as I will argue in the next chapter, not all) of the practitioners of computer-generated poetry I worked with are content with delegating this exploration to the computer, whose randomizing function creates the appearance of nondeterministic creativity and of deviation from the constraints that they fashion as a kind of pseudo-clinamen. Their insistence on not altering or editing the output texts of their poetry generators demonstrates how the use of a computer in a literary strand that traces its roots directly to the Oulipo has helped to swing the pendulum back to the kind of aesthetic principles formulated by the Oulipo's first generation of writers, who adhered to strict proceduralism and downplayed the task of exploring the implications of their procedures as a kind of "applied work" of a lesser value.

Another prediction made by members of the Oulipo apropos the use of computers in poetry becomes similarly problematic in light of the important role that randomization plays in computer-generated poetry:

> It is because of the *potential* the computer furnishes to the mechanistic model that the Oulipians are drawn to it. The computer constitutes thus another arm in the arsenal they deploy against the notion of inspiration and, in a broader sense, against the avowed *bête noire* of the Oulipo: the aleatory. For another way of considering the Oulipian enterprise is as a sustained attack on the aleatory in literature... All of their work... may be read in this light. As Jacques Bens expresses the position: "The members of the Oulipo have never hidden their abhorrence of the aleatory, of bogus fortunetellers and penny-ante lotteries: 'The Oulipo is anti-chance,' the Oulipian Claude Berge affirmed one day with a straight face, which leaves no doubt about our aversion to the dice shaker. Make no mistake about it: potentiality is uncertain, but not a matter

of chance. We know perfectly well everything that can happen, but we don't know whether it will happen." (Motte 1998:17)

The background to the Oulipo's aversion to chance and the aleatory is Queneau's opposition to everything surrealist. Queneau, who in the 1920s was a member of the surrealist movement and who later broke away from it, envisioned the Oulipo as "a conscious antithesis of Surrealism" (Duncan 2019:83) in terms of both its institutional organization and its aesthetic principles. With respect to those principles, Queneau focused his criticism on the surrealists' writing methods, especially automatic writing, in which the writer forgoes any conscious intention and instead writes automatically everything that comes to his or her mind, thereby tapping into his or her unconscious. Queneau argued that inspiration functions in this writing method as "something which strikes at random, and which the artist waits for passively, hopefully" (Duncan 2019:85), as well as that "the poet who pretends to 'dive' into his unconscious to bring up the marvels and new worlds announced by Apollinaire is not an experimenter but an empiricist. He waits open-mouthed for inspiration like an entomologist hoping to catch an insect" (83–84). Automatic writing establishes "the equivalence . . . between inspiration, exploration of the subconscious and liberation, between chance, automatism and freedom. Now, this inspiration which consists in blindly obeying every impulse is in reality a slavery. The classical writer who writes his tragedy observing a certain number of rules that he is familiar with is freer than the poet who writes whatever comes into his head and who is a slave to other rules which he ignores" (85–86). In this Oulipian framework, "the writer—properly trained in the craft of writing—should *always* be able to write" (85), "the only literature is voluntary literature," and "the really inspired person is never inspired, but always inspired" (Lescure 1998:34). The Oulipo's goal was thus to force "inspiration . . . [to] cease to be capricious" and to make it a property "that everybody might find . . . fruitful and compliant to his desire," a property purged of the "romantic effusions and the exaltation of subjectivity" by which literature as a whole had been dominated (34).[8]

A writing method that the Oulipo devised and explicitly contrasted with automatic writing is the method known as S + 7, which "involves taking an existing text and replacing every noun . . . with the seventh noun following it in a given dictionary" (Duncan 2019:86). This writing method, argued Queneau, is "a new kind of automatic writing: rule-governed, rather than rule-free, but nonetheless markedly simpler to implement than its predecessor" (88). In this technique, "human agency is expunged almost entirely: 'the operator is reduced to a purely mechanical function'" (86). Crucially, Queneau argued, its "results are ostensibly as before—'Surrealist, or at least seemingly Surrealist'" (89), and this similarity of results undermines any added literary value that automatic writing may have by virtue of its presumed ability to give the author access to his or her unconscious. Whereas the surrealists prioritized the Romantic notion of the creative artist, the Oulipo's procedural writing methods demonstrated the primacy of the unconscious as language.

The problem with these Oulipian arguments is that while some of their procedures consisted of "arithmetical series that imitate chance while obeying a law" (Roubaud 1998:88), many of their other procedures incorporated different forms of chance. For example, as Queneau himself acknowledged, "the result [of the S + 7 writing method] obviously depends on the dictionary one chooses" (quoted in Duncan 2019:86). Choosing this or that dictionary is not that different from sorting out a list of verbs alphabetically, which exposes the randomness of "how things are," as Jason and the participants discussed in the workshop in relation to Gregory's computer-generated poem, discussed in the previous chapter. In addition, deciding to use the number seven rather than six is not that different from throwing a die each time one encounters a noun in a text in order to decide with which noun in a dictionary one should replace it. The Oulipo's alleged "aversion to the dice shaker" (Motte 1998:17) should thus be significantly qualified, as should critical accounts that argue that the group "favored writing under programmatic constraint instead of chance operation" (Funkhouser 2007:33).

While acknowledging the aleatory component of some of their constraints, members of the Oulipo argued that their writing methods were more rational than the surrealists' method of automatic writing for two reasons. First, "writing under constraint is superior to other forms insofar as it freely furnishes its own code" (Bénabou 1998:41). This holds true for constraints that involve a random element, too. Even if a specific constraint results in nonregularity, "the nonregularity is not accidental: it results from the decision to use it, thus is predetermined, thus is constrained" (Roubaud 1998:88).[9] Second, the Oulipian writer crafts very specific forms of randomization, rather than general and vague ones. It is the specificity of these forms (for example, replacing every noun in a text with the seventh rather than sixth noun following it in a given dictionary) that gives members of the Oulipo the justification for arguing that although the potentiality produced by their procedures "is uncertain," it is "not a matter of chance. We know perfectly well everything that can happen," as opposed to automatic writing. Rather than being anti-chance, then, members of the Oulipo present themselves as anti–forms of chance that they did not intentionally decide to use or craft and that are not based on clearly defined rules and procedures.

However, even these qualifications are problematic. First, members of the Oulipo frequently "know perfectly well everything that can happen" as a result of their self-fashioned constraints only in theory but not in practice. What does it mean to know everything that might happen in relation to a constraint that is generative of a hundred thousand billion poems, as is the case with Queneau's archetypal poem? Here a constraint's immense potentiality is seen to be a double-edged sword. On the one hand, this poem is touted for the way in which its very specific rule-governed architecture is responsible for its enormously fecund literary potential. On the other hand, this potential means that despite the poem's rule-governed architecture, one cannot pretend in any meaningful way to "know perfectly well everything that can happen." All that the Oulipians know for sure about this and similar kinds of poems is what the procedure that is responsible for "everything that can happen" is, rather than "everything that can happen" as a result of following this procedure. To really know everything that can happen, they would need to follow a poem's procedure and see how it unfolds, which in the case of Queneau's poem would take many lifetimes.

The second and more important problem concerns the argument that because it is the writer who fashions the constraints with which he or she works, he or she is more in control of his or her writing process, even if those constraints incorporate chance and randomness. The form of computer-generated poetry that was created and discussed in the workshop casts doubt on this argument because it represents a form of relinquishing control in two ways. First, "very few people are able to just look at a code and know immediately or accurately what the output will look like," as Sid told me in the vignette I discussed in the previous chapter. If, because of code's abstraction, the poet frequently needs to execute the code in order to understand what its implications are, then the level of control the poet has is already reduced. Second, if to the significant gap that exists between writing code and knowing what its implications are is added the fact that this code frequently incorporates computer-based randomization (Funkhouser 2007:18; see also 79, 83), then the poet's ability to conduct informed experiments with literary constraints might not be as obvious as it first appears to be. Gregory's description of the computer as a kind of inscrutable and independent creative agency that frequently exceeds and subverts his authorial intentions, which I discussed in the previous chapter, suggests that the literary "experimentalism" of some of the people I worked with is often experimental in a sense that brings them a little closer to the surrealists' experiments, as the game of chance with which Jason opened the workshop also suggests. This is not to deny that the workshop participants were active and intentional: they devised different constraints and wrote code and tweaked it based on the output texts that were generated when the computer executed their code. However, because of their limited ability to conduct rigorously informed experiments with code, some of them, too, for all intents and purposes, were not completely far removed from the surrealist, for whom, according to Queneau, inspiration is something that "strikes at *random*, and which the artist waits for passively, hopefully" (Duncan 2019:85, emphasis added). In other words, although my interlocutors saw themselves as "ant-makers," to use Jason's felicitous reference to Herbert Simon's argument, which I quoted in the second of the two vignettes that open this book, and were thus certainly different from the surrealists who, according to Queneau, waited "open-mouthed for inspiration like an entomologist hoping to catch an insect" (Duncan 2019:83–84), the fact that many of the "ants" my interlocutors "made" were based on different kinds of computational indeterminacy brought them closer to operating like an entomologist nonetheless.

In summary, the analysis of the continuity and change between computer-generated poetry and the Oulipo puts the aesthetic dimensions, tensions, and contradictions that structure computer-generated poetry in their historical-intellectual context. It reveals what is new about computer-generated poetry in relation to the Oulipo, what the former shares with the latter, how the former problematizes the latter, and how the latter problematizes the former.

As I argue in the next chapter, some of the practitioners of computer-generated poetry with whom I worked expressed their reservations about this art form's aesthetic dimensions and textual outputs in relation to the tensions and contradictions that I explored in this chapter, especially those that result from what they experienced as the prioritization of technological and computational-algorithmic means over their poetic ends. In doing so, they also expressed their desire to go back to and reinhabit the aesthetic dimensions and textual outputs that they associated with the kind of nondigital poetry they had previously felt to be in need of digital transformation.

8

Crosscurrents and Opposing Perspectives

Two Moments of Reified Computation

I was about to make my way to the workshop's location for the first time. Although it was the beginning of what would be a weeklong heat wave, I decided to walk from my hotel to the workshop instead of taking the subway, thinking that my past experience with Israel's brutally hot and humid summers had surely prepared me for whatever New York's could throw at me. By the time I arrived at the building complex where the workshop was to take place, however, I regretted my decision. Completely exhausted, I entered the complex's courtyard and rushed to the performance space in which the workshop was held, hoping to find some refuge in what I had imagined would be a room comfortably cooled down by benevolent air conditioners. That the space's door was ajar did not bode well. I entered the space, and an onslaught of heat immediately discombobulated me. Inside, a few of the participants were already sitting around two tables, as visibly exhausted and dazed as I must have appeared to them. After all the participants had arrived, Tim, a man in his thirties who was one of the cofounders of the organization that put the workshop together, introduced himself and said:

> This building is a very old building. . . . Funny enough, there used to be a research center here decades ago. And they did a lot of scientific research here, including Claude Shannon. And there is a story about how Alan Turing came here to visit Claude Shannon and they discussed whether machines could ever play chess, if machines could think, which is a very interesting question to us. . . . And this place was designated as a historical monument by NYC, which is great but it means we cannot adjust the space for our needs. So the place is really raw. This is why we cannot install an AC outside of the building. You are

probably wondering why it is so hot in here—that is the reason. [Tim wipes the sweat off his forehead and adds apologetically:] So it is part of the experience I am asking you to endure in the next few days.

*

An hour and a half before the workshop's concluding event, in which each of the participants would present his or her work, Jason and some of the participants were discussing the order of presentation. Jason said: "We need to decide on the order. There are fifteen presenters. Leandra, choose a number between one and fifteen." Leandra, one of the participants, looked at Jason and said: "I don't know which number to choose!" "You know what," Jason said, "let's write a short program using the Commodore's random number generator to create the order of presentation." Jason and the participants approached the Commodore 64, which Jason had brought with him to the workshop and which throughout the workshop had stood on a table in the side of the room. When I asked Jason on a separate occasion why he had brought this computer to the workshop, he responded that the Commodore 64 is "a classic, a well-designed system for these short brief explorations into computing. And it's very welcoming. You turn it on, you program. You can write a one-line, two-, or three-line basic program." Jason tried to turn the computer on, but for some reason he couldn't. After fiddling with it for a few minutes, he told the participants: "I can use the Commodore's software on another computer." "No!" Leandra objected. "That would mean that we can do it with any computer rather than with this one." "Yes, but it is not working, so we have no choice," said Jason. He took his laptop and started coding quickly. After a few minutes, he ran the program. The participants looked at his screen. "It doesn't work," Jason said with disappointment; "I don't know why." Gregory commented: "So much effort just to produce random numbers!" Jason made a few changes in the code, which then worked. "Oh," Jason said, "it will only give us numbers between 0 and 1, so we have to divide 1 by 15 because there are fifteen participants." After making some calculations and fiddling about with the code, he gave Leandra a number and then proceeded to do so for the rest of the participants, thus establishing the order of presentation for the workshop's looming concluding event. Leandra declared: "It was totally worth it!"

*

These two vignettes, the first opening the workshop, the second closing it, turn on the vastly different ways in which digital computation can be transformed from an enabling means to a limiting end. In the first, the physical location in which some of the foundational people in the history of digital computation—Claude Shannon, Alan Turing—researched and debated the possibility of artificial intelligence and whether computers could think, is designated a historical monument by the city. The result of this decision is that the present users of this location, who would like to experiment with the design and use of digital computation for poetic purposes, are required to work in quasi-prehistoric conditions that force them to face the most "analog" material and embodied difficulties, such as heat, sweat, and physical exhaustion. In the second, computer-based randomization, which my interlocutors in all three ethnographic sites prized because it allowed them to open up new creative spaces, becomes an obstacle that exhausts their creative energies. Although their attempts to use digital computation to generate a random order of presentation are thwarted time and again, they keep insisting on trying to do so instead of finding other ways of establishing the order. Furthermore, they insist not only on using digital computation for this purpose, but on doing so on a very specific material artifact that has a place of honor in the history of digital computation, the Commodore 64. Inasmuch as the Commodore 64 has become a kind of venerated computational relic, its transformation from an enabling means to a limiting end represents the same kind of transformation of the workshop's historic location from a place in which some of the most revolutionary developments in digital computation took place in the past to a place that limits people's ability to find relief from a hot summer day in order to realize the creative potential of digital computation in the present.

The transformation of digital computation from an enabling means to a limiting end is at the core of the reservations that some of the workshop participants expressed about the kind of computer-generated poetry taught and practiced in the workshop. In chapter 6 I discussed one focus of those reservations, namely the notion that text distribution and linguistic texture by themselves can be sufficient aesthetic goals. In the present chapter I first discuss two additional foci of those reservations: the idea that precision of coded form or constraint is aesthetically important in and of itself, and that such a form or constraint can aid the poet in his or her writing process. I then explore how the poetic practices of some of the participants and of one of the guest presenters during the workshop subverted these ideas, which can be traced back to the literary debates between members of the first and second generations of the Oulipo, and represented a search for complementing aesthetic goals and rationales.

Aesthetic Questions and Doubts

PRECISION OF CODED FORM

On the second day of the workshop, the following exchange took place after Jason presented a classic computer-generated poem:

LAURA: I think of poetry as a very precise form of art, but generative poetry seems to be more fluid and not precise enough.

JASON: Let's just talk about what form is rather than what poetry is. Whether a poem must have form—from my perspective, certainly. I write poems in forms. They are forms that I, for the most part, invent, or that are user-constrained and that I adapt and adopt in various ways. And so this is a poem in form. And the form actually is right here. The code is describing what it is. . . . It's generated by a one-page stand-alone Python code that's in the back. So form is important to me. . . . Form is important and these [poems] are not less formal than a sonnet or a haiku.

Laura argues that it is hard for her to reconcile computer-generated poetry with how she typically thinks about poetry. She suggests that the nondigital poetry she is used to is an art form in which the poet chooses every word very carefully, whereas computer-generated poetry employs repetition and redundancy to create textual patterns that do not seem to reflect such a careful choice. In response, Jason argues that computer-generated poetry is also a very precise form of art. However, its precision is the precision with which the poet writes code, and the precision of the form that the code "describes" and that the computer then realizes. Jason's response shifts the locus of precision from the level of the words that might appear in any output text of a poetry generator to that of the code he or she writes when creating this generator.[1]

Jason's emphasis on the importance of the precision of form mirrors the early Oulipo's emphasis on the importance of rigorously developed and executed formal constraints. For example, Jacques Roubaud, a member of the Oulipo, articulates a series of propositions that are supposed to underlie Queneau's poetics, among which are "to comport oneself toward language as if the latter could be mathematized"; "if language may be manipulated by the mathematician this is because it may be arithmeticized;" "arithmetic applied to language gives rise to texts"; and "language producing texts gives rise to arithmetic" (Roubaud 1998:82). This highly formalized framework results in further propositions, such as: "writing under Oulipian constraint is the literary equivalent of the drafting of a mathematical text, which may be formalized according to the axiomatic method" (89). This means that "a constraint having been defined, a small number of texts (only one, in some cases) are composed by deduction from this axiom" (91).[2] In all of these statements the precision of form and its realization by "deduction" take center stage, but the meaning of the resulting poetic texts is not discussed at all.

Some workshop participants pointed to inconsistencies in the arguments made by the facilitators in support of precision of form or constraint as a satisfying basis for computer-generated poetry. For example, on the workshop's third day, Jorge, a guest lecturer in his mid-thirties, told the participants: "I wouldn't say anything about what literature is. If someone tries to tell you what literature is, you should run away from this person. What I will do is tell you how I like to think about literature and about digital literature. So this is a famous passage from Roland Barthes, from his famous text 'The Death of the Author.'" Jorge screened a page from Barthes's essay. "At the end he says this," Jorge said while reading the following screened text: "Thus is revealed the total existence of writing: a text is made of multiple writings, drawn from many cultures and entering into mutual relations of dialogue, parody, contestation, but there is one place where this multiplicity is focused and that place is the reader, not, as was hitherto said, the author. The reader is the space on which all the quotations that make up a writing are inscribed without any of them being lost; a text's unity lies not in its origin but in its destination" (Barthes 1977:148). Jorge then continued: "The other day we were talking about the function of the reader.[3] So the first thing I wanted to say is that, as we were seeing, most of the time it doesn't matter much what the author wanted to say but rather what you can do with that [as a reader]. And with these texts that we are using here, it is particularly important. So this is just a way I like of thinking about literature."

In itself, Jorge's invocation of Barthes is another indication that the workshop facilitators focused on a strand of computer-generated poetry that resonates with the "syntactic" rather than the "semantic" Oulipo, that is, with the group's first rather than second generation of writers. Barthes's ideas about literature and language were part and parcel of the structuralist zeitgeist that influenced the Oulipo in its early days, especially the emphasis on language as a concrete object that exceeds whatever intentions the author has and whatever conscious or unconscious meaning the author wants or is compelled to convey (Duncan 2019:90–91). At stake is a kind of "agency without authorship" (94).[4] Barthes's ideas also aligned more specifically with the attack made by the Oulipo's first generation of writers on surrealism's fetishization of the author and his or her unconscious as a source of poetic ideas, as I discussed briefly in chapter 6. His ideas provided them the conceptual machinery with which to accommodate and even celebrate the kind of nonsensical literary output that a computer might produce and that some of the workshop participants thought was problematic. Thus, when the French cyberneticist Albert Ducrocq built an electronic poet in the mid-1950s, which he called Calliope (after the Muse of epic poetry), Boris Vian, a novelist and close friend of Queneau, argued that "a machine like Calliope would always have the upper hand in producing certain registers of poetry" (Duncan 2019:31). More specifically, "in the realm of the vague, the bizarre, the ephemeral, the abstruse and the dream-like, the robot will beat us every time. Effectively, it will have none of the bad reasons for choosing this or that term which our past imposes on us" (ibid.). This past includes both "literary influence and its anxieties," as well as the poet's own authorial subjectivity and "set of experiences that [he or she] might want to express" (ibid.). Inasmuch as this electronic poet's "method is simply combinatorics, and its materials are whatever vocabulary it has been fed . . . [its] poems raise the possibility of the text voided [*sic*] of intention, free of authorial subjectivity, an idea which will find its theoretical apotheosis fifteen years later in Barthes's famous 'Death of the Author' essay" (31–32). The Oulipo's first generation of writers thus viewed the electronic poet's nonsensical literary products ("Ducrocq clarified that only about one in four of Calliope's prose poems was actually intelligible" [30]) as a feature rather than a bug, because those products were aligned with the overall philosophical and aesthetic framework they promoted.

However, Barthes's essay focuses predominantly on dismantling the idea of the intentional author as the locus of a text's meaning, that is, the author who consciously informs his or her writing by specific intentions and thinks that those intentions can be discerned in his or her text. Barthes argues that the multiplicity of relations and meanings that underlie any use of language subverts whatever intentions an author has in writing a text, and that when a text reaches the reader it is the reader, based on his or her unique makeup, who gives it whatever coherence and unity it may have. In contrast, Jorge's use of Barthes's argument in the context of the workshop puts the burden of meaning-making on the reader, although computer-generated poetry's output texts, as opposed to the texts discussed by Barthes, frequently make very little sense from the very beginning, that is, even to its "authors," as the workshop participants observed frequently.[5]

Participants' reactions to Jorge's explanations suggest that they did not find them entirely satisfactory. For example, later in the day Jorge introduced a few programs he had written in the past. He then asked the participants to experiment with them: "When I was working with Jason in the past year and a half," he said, "we created very tiny programs. We focused on the size of the program. So I'll show you some of the programs. For instance, what we saw before was 256 bytes." Jorge showed the participants the code of the poetry generator whose output text he had read before. "This is the program," he said. After showing the participants how to retrieve the code from his website, Jorge asked them, "Do you want to test this a little bit, to play with it?" The participants experimented with the code for five minutes. Jorge then asked: "Did anyone get any interesting results? It was only for you to understand how it works, but perhaps you have something to share?" A number of the participants read their results. Almost all of them were nonsensical, as the following result read by Allen, a participant, demonstrates: "That I did that I am did I will and but that I am but that I am not that I'm perhaps. . . ." After a short discussion, Gregory said to Jorge: "Earlier you said that the unity of the text is not in the source but rather in the destination. But here, because the goal of the program is to be small, the meaning of the [resulting] text is so obfuscated—it is very hard to understand it. It doesn't express anything. So in this case the unity is in the source—the code—and not in the destination [i.e., the reader]." Gregory points to a problem in Jorge's appropriation of Barthes's ideas. Faced with the nonsensical output texts produced by the code Jorge presented, Gregory argues that the unity of the text cannot lie in the reader, for no reader would be able to find meaning in or give meaning to such an output text. Rather, the unity is in the specific constraint that underlies the poetry generator that produced this text: the stipulation that its program should be "tiny" and consist of only 256 bytes. Gregory argues that Jorge's poetry generator represents an emphasis on the constraint, form, and code rather than on whatever output texts they might give rise to and hence, in some ways, at the expense of the reader (cf. Tomasula 2012:486).

Here it might be useful to return to the second of the two vignettes with which I opened this book, in which Jason invoked Herbert Simon's argument about the basis of the complexity of an ant's convoluted path on the beach, and which Jorge alluded to after he presented Barthes's text in order to point to the importance of the reader. Whereas in that vignette Jason argues that a poetry generator is akin to an ant, in that it is very simple, and that the complexity of its output texts is the result of the meaning that the reader finds in those texts based on his or her cultural background and associations, Gregory's comments suggest that writers of computer-generated poetry (and their predecessors, such as the members of the Oulipo's first generation) might risk constructing ants with no beach to walk on, as it were; that is, they focus on the precision of coded (or mathematically formulated) form as an end in itself rather than as a means to producing satisfying poetic ends. In so doing,

CROSSCURRENTS AND OPPOSING PERSPECTIVES   203

they might create output texts that few readers can relate to, find meaning in, and imbue with complexity.

Inasmuch as the facilitators' stance resonates with the kind of arguments that the Oulipo's first generation of writers espoused, the inconsistencies that some of the workshop participants found in that stance problematize those earlier arguments, too. Gregory's objection to Jorge's appropriation of Barthes's ideas points to what was perhaps a kind of false consciousness among the Oulipo's first generation of writers, who, while focusing their energy on the construction of mathematically informed literary constraints, also argued that "the role of the reader is . . . capitally important" and that "potential literature would be that which awaits a reader, which yearns for him, which needs him in order to fully realize itself" (Bens 1998:66). It also problematizes those literary critics who accept at face value that the "Oulipo's early period" was characterized by "reader-oriented, highly structuralist tendencies" (Duncan 2019:101). Rather, the Oulipo began in earnest to be more reader oriented only when its members agreed to break free from the fetishization of, and strict adherence to, mathematically informed and rigorously followed constraints.

DIGITAL CONSTRAINTS AS A SOLUTION AND A PROBLEM

The workshop facilitators offered another rationale for computer-generated poetry, one that focused on the writing process of this art form rather than on the aesthetics of its literary products or the precision of its coded form. In an exchange later on the workshop's third day, Jorge told the participants:

> Earlier I told you that when I was working with Jason in the past year and a half, we created very tiny programs. The idea of having this constraint was—a month ago I was at this conference and someone was talking about that difference that Walter Benjamin had pointed to in "The Work of Art in the Age of Mechanical Reproduction" [Benjamin 1969]. The difference is that at that moment [in 1935, when Benjamin wrote his essay], production was very expensive. It was hard to produce things. But now we have the opposite problem, this artist was saying [at the conference]. We have an abundance of options. Everything that we do creates immense possible outcomes. And maybe a way of dealing with that is having some constraints. And the main concept of the Oulipo was constraints. The reason they gave was different. In their case they were opposing the surrealists who were too irrational. So they wanted to create rational constraints. [After discussing Queneau's *One Hundred Thousand Billion poems*, Jorge continues:] So, as you can see, with very few elements you

can make a very big thing. So before you continue with the code, I want to mention other constraints that were used by the Oulipo.

Jorge continued to describe the S + 7 writing method (which I discussed in the previous chapter), the Snowball method (in which each line in a poem has only one word and each word must be one letter longer than the word in the previous line), the lipogram, and palindromes.

Jorge draws a clear parallel between the Oulipo and the computer-generated poetry that was the focus of the workshop: both are based on writing poetry under self-fashioned rational constraints. In addition, he explains the necessity of having constraints by invoking Walter Benjamin's famous essay. Based on a conference presentation that he attended, Jorge argues that in the contemporary moment, when production is so easy—especially given the availability of digital technologies—constraints have become more necessary than ever as a way for artists to orient themselves and focus their creative energies. Later that day, Jorge added:

> You just need one [constraint] to write something. You can see a constraint as a rule, and you can use a rule in a constructive way when you create a set. So in this case I think that reducing the size of the program—the tiny programs Jason and I wrote—makes you stop thinking about things that are useless and even impossible. In different circumstances there are things that you'd like to have in your outcome, but this is not essential to my program. And in that sense it is very interesting. I am not saying that you must have constraints to achieve anything, but it's just a way to achieve a result and to get a starting point.

Although Jorge argues that the members of the Oulipo gave a different rationale for using constraints than the rationale given by the conference presenter, the Oulipo writers were not oblivious to the argument, voiced by Jorge, that "we have an abundance of options. Everything that we do creates immense possible outcomes. And maybe a way of dealing with that is having some constraints." For example, Perec suggested that "writing a poem 'freely' would be more problematical for him than writing according to a system of formal constraint: 'I don't for the moment intend to write poetry other than in adopting such constraints. . . . The intense difficulty posed by this sort of production . . . palls in comparison to the terror I would feel in writing 'poetry' freely'" (Motte 1998:13). Jorge's point that a constraint can provide the writer with a good "starting point" mirrors the arguments made by other members of the Oulipo, namely that constraints can help writers "in finding the springboard of their action" (Lescure 1998:38), and that, "rather than stifling the imagination," they can serve "to awaken it" (42). Similarly, Jorge's suggestion that "reducing the size of the program—the tiny programs Jason and I wrote—makes you stop thinking about things that are useless and even impossible" mirrors the Oulipian argument that "the choice of a linguistic constraint allows one to skirt, or to ignore, all these other constraints which do not belong to language and which escape from our emprise" (Lescure 1998:42-43).

One problem with this justification for using self-fashioned constraints is that it merely shifts the burden of choice and endless possibilities from the act of writing to the act of fashioning (or choosing) constraints of writing. It is not necessarily the case that starting with the idea of constraints can better restrict and focus literary production, because one can in principle create an unlimited number of different constraints. This is especially the case with respect to the work of art in the age of digital production and reproduction, to riff on the title of Benjamin's essay. In this digital computational age, not only do constraints not offer a clear solution to the problem of too many possibilities and courses of action, they can become a new dimension of this problem, for a number of reasons. First and foremost, even for a person with basic knowledge of programming, it is fairly easy to fashion constraints in the form of writing code and then to explore these constraints by having a computer execute the code. As Jason told the participants on the workshop's first day: "We are working very hard to provide you with a kind of guitar. You know what they say about a guitar: it's low stairs, high ceiling. You can make sounds that are pretty good from the very beginning without knowing much about it, and you can keep going and you can become a virtuoso and it has lots and lots of possibilities." Although Jason views the ease and facility with which people who are new to coding can fashion productive constraints as an advantage, that means that the act of fashioning constraints can be beset by the same problem of an abundance of possibilities, the very problem it was supposed to solve, at least according to Jorge.

More important, digital computation makes it very easy to change constraints in numerous ways by means of editing the code that underlies them. The fact that it takes very little effort to change a poetry generator's code and see the implications of those changes increases the temptation to continue to do so in the hope of reaching more interesting results. In a conversation I had with Gregory he commented on this aspect with respect to his work as a programmer in a design firm:

GREGORY: Deadlines are crucial for me. I set myself deadlines. Or I get deadlines set by clients. My best work resulted from a set deadline because otherwise there are too many possibilities that you want to explore. You can do anything with computation. Sometimes even if I am happy with the results, I might alter something in the system that will change the whole thing.

EITAN: Why would you do that?

GREGORY: Well, because there *are* so many possibilities! I think to myself, "Oh, this might be interesting. Or that might be interesting." Like, you cannot un-chip marble, but with digital literature you can go back and forth all the time.

Gregory's argument that "you cannot un-chip marble" refers to a comparison he had made elsewhere between digital art and sculpture. A sculptor who sculpts with marble begins with a given block of marble that has specific characteristics limiting what he can do. The sculptor then "chips away until a form emerges," as Gregory put it, and this chipping away is an irreversible process. In contrast, the coding artist "starts with a blank canvas." He creates "the system that creates" the resulting art. The core dimension of making this system is writing "rules—algorithms written in your language of choice that make processors churn to your will." He elaborated:

> You add layers of code until a system is set up. At this point your system can produce a large, but finite set of images. Anything inside the system is possible, anything outside of it is not. Think of this as your possibility space, your block of marble.... So first you make your marble. Once the rules are in place, the next step is to configure your system—picking the values.... Changing the [values] does not change the underlying system, but it has a fundamental impact on the output.... By changing various values, you explore the [art] that your system can produce, in search of the [art] you want it to produce.

This stage, Gregory argued, can be compared to the stage of the sculptor's chipping away at the marble. "Of course," he cautioned, "the break is rarely as clean as this. You often go back and augment the system by changing rules, because they are not actually set in stone. It's more like clay. Or, really, it is whatever you want it to be, or can make it to be."

Gregory's thought-provoking comparison highlights three crucial differences between digital art and sculpture. First, in digital art the artist creates the space of possibilities that he can then explore with the help of the computer—that is, he creates his own block of marble.[6] Second, the process of exploring this space of possibilities is reversible, unlike the process of chipping away at a block of marble. The artist can change the values in the rules that he crafted again and again until a satisfying result is produced. Third, the artist can even go back and change pretty easily the space of possibilities that he first created ("changing rules") and thus create a new space, rather than continue with the first space of possibilities that he created (cf. McLean and Wiggins 2012:246–247). In contrast, a sculptor has to either continue to work with the block of marble that he started with or make the costly decision to replace it with a completely new one and start again from scratch.[7] Gregory's comparison provides a rich context in which to understand why, in contrast to Jorge's argument, some of the people I worked with did not think of constraints as a solution to the problem of overproduction and the abundance of possibilities that they face when they want to exercise their creative agency in the context of digital art, but rather as a new dimension and determinant of this problem.

Complementing Aesthetic Goals and Rationales

Some of the workshop participants expressed their desire to go back to and reinhabit the aesthetic dimensions and textual outputs that they associated with the kind of nondigital poetry they had previously felt to be in need of digital transformation. They did so against the backdrop of their reservations about computer-generated poetry's aesthetic dimensions and textual outputs in relation to such tensions and contradictions, especially those tensions and contradictions that result from what they experienced as the prioritization of technological and computational-algorithmic means over their poetic ends. Their search for complementing aesthetic goals and rationales took three main forms.

IN SEARCH OF THE SINGULAR POEM

Workshop participants occasionally expressed their desire to reexperience the singular output text or textual construction that are the foundations of traditional, nondigital poetry. They did so as a result of their discomfort with linguistic texture and with precision of coded form as sufficient rationales for computer-generated poetry. Their stance differed from the aesthetic ideals that most of the workshop facilitators encouraged them to inhabit. In the context of these ideals, one should approach a poetry generator's output text not as a singular entity, but rather as an indication (or "sample") of the general text distribution that the poet defined by means of code. Approaching one of a poetry generator's potential output texts or a stretch thereof as a singular entity would make little sense. Thus, when I asked Jason about the possibility that specific output texts might have a singular aesthetic value, he responded by using the same polling imagery that informed his responses to many of my other questions: "I think the perspective on it is—again—it's sort of like asking—it's like asking someone conducting a Gallup poll, and they say, 'Yeah, we talked to I don't know how many people in today's population . . . 2500 Americans.' So they do their polling and they talk to 2500 people or whatever and, then, why would it be sensible for someone to tell them, 'Oh, you really should talk to this specific guy'? [Jason laughs.] I mean, it doesn't make any sense, right?" Jason argues that to cling to a specific output text of a poetry generator—whether an already produced text or a future potential text—would not make any sense because of the text-distributional aesthetics that informs this kind of poetry. In the same way that a statistician conducting a poll about the American population would not be interested in talking to any specific American person, it would not make much sense for the poet to be interested in a specific output text or stretch thereof of his or her poetry generator except as a token of the general type that is the text distribution the generator can produce.

Although some workshop participants were able to embrace this text-distributional aesthetics, other participants preferred to find in the output texts of their poetry generators the kind of textual singularity that is associated with traditional, nondigital poetry. In other words, they behaved precisely as pollsters who are interested in "talking to this specific guy." One way they expressed this desire is by refusing to instruct the computer to execute their code again, so as not to lose a specific output text that had just been generated. For example, on the workshop's fifth day, Laura ran the program she had written based on a prompt given in the previous session. The program incorporated randomness, and she would run it, read the results aloud, and then run it again. After four iterations, Chris stopped her and asked the participants for their comments. Sandra said: "At one point the way you read it sounded like . . . a poem that I would encounter in the real world. And when you hit it to run again, part of me was like: 'No!'" Sandra's reaction is noteworthy for two reasons. First, Sandra values the specific output text that Laura read because it resembles "a poem I would encounter in the real world," that is, a poem that is not informed by the semantic incoherence characteristic of so many computer-generated poems. This aspect of her reaction mirrors the fact that as the game of chance, which I discussed at the beginning of the previous chapter, progressed, participants showed appreciation only for those question-and-answer pairs that made simple sense even though they were the product of a chance procedure. Second, Sandra prefers a specific output text of Laura's poetry generator over its other output texts and expresses her sadness at realizing that once Laura runs the program again this specific output text will be lost. Similarly, on the workshop's fourth day, Sarah ran the code of a specific poetry generator that she had created and then read its output texts to the participants who could see them on the television screen. Sarah commented: "This all seems to work together in the context of a country fair, in some way. So I'm not going to run it again because I want to keep this. If I run it again it will go away."

The desire to experience this form of textual specificity could also be gleaned from the fact that some participants intentionally continued to execute a poetry generator's code time after time in the hope of eventually generating an output text with singular aesthetic value. In the statistical polling imagery Jason used to answer my questions, the decision to continue to explore a poetry generator's textual implications with the hope that a specific aesthetically singular output text might be produced makes little sense, too. A poetry generator's past, present, and future output texts will remain essentially equal to one another as long as its code remains unchanged. They are all snapshots or samples of the same textual pattern or linguistic texture or distribution of text. Thus although, according to scholars of computer-generated poetry, "computers cannot be programmed to engineer a 'perfect' poem" (Funkhouser 2007:79) that will "remain in your memory as something unique and perfect" (Balpe 2003:7, quoted in Funkhouser 2007:83), nor should they be expected to do so (80), some workshop participants hoped to find "perfect poems" in the output texts generated by their poetry generators. They displayed a desire to experience and inhabit the singularity of specific output texts precisely because of the fact that the computer-generated poetry they experimented with represents not only a kind of formal abstraction on the level of coded form, but often also textual abundance and excess on the level of the output texts that result from realizing this form.

THE PUNCTUM OF COMPUTER-GENERATED POETRY

The formal abstraction and textual abundance and excess that characterize computer-generated poetry can also lead its practitioners to reappreciate the singularity of the nondigital texts they occasionally use as raw material for their generators to process. For example, on the workshop's fourth day, Sarah told the participants:

> Here is something a little more complicated that I did, which involves—still working with my own work, but also coming up with the possibility of working with different kinds of work, work that is not my own written work. So one thing I did was to take texts from Project Gutenberg: [Walt] Whitman's *Leaves of Grass* and *Goblin Market* by Christina Rossetti, which is an excellent poem. . . . So essentially what I'm doing is I'm bringing in my work and the sample poems documents. It's [the computer program] randomly picking a

number of words from a line, a random sample from the words, and joining them together. It's a little more complicated. And what I found doing this, I did it in a class where we used *Leaves of Grass* and [Edgar Allen] Poe's *The Raven*, and one of the words that appeared was "llama," and one of the students was like: "This can't be Whitman because there are no llamas in America." And I was like: "Well, it doesn't mean that Whitman couldn't have put 'llama' in his poem." So we went back and we looked at the source texts and we discovered exactly where llamas appeared, and also where it always appeared in our text, too, which turned out to be very interesting. So you see and learn interesting things when you work with texts of your own or somebody else's.

The appearance of a specific word (Whitman used the spelling "llama" for "lama") in an output text of a poetry generator that uses Walt Whitman's classic poem as raw material motivated Sarah and her students to go back to *Leaves of Grass* and to discover where that word appears in it.[8] In doing so, Sarah and her students refused to remain on the poetry generator's text-distributional level. They chose instead to zoom in on the micro-level of one particular word in one of the generator's output texts, as well as on this word's location in the nondigital poem that provided the raw material for the code of the poetry generator to process.

An even more powerful reappreciation for the singularity of the nondigital texts that some workshop participants chose as raw material for their poetry generators to process resulted when those texts were the participants' own previously written nondigital poems. Thus, on the same day Sarah showed the participants a poetry generator that uses her own corpus of nondigital poems as raw material. Rather than leaving intact the generator's output texts, her intention was to mine them for new material that she could use when writing new nondigital poems. After running the program, she started reading the results and said:

> Oh, I like this! "You never tongued above me." That's great! You know, I don't know where these lines actually come from in my work. . . . These are texts that I've written in the past but I couldn't tell you where these phrases come from. . . . But they are presented now in a new way: "Oh, you never tongued above me." . . . It's in a previous work that I've written, somewhere in there. [Sarah pauses for a few seconds and then says:] Actually, now I think I just remembered where it comes from. And that's kind of cool that I can look at it—"you never tongued"—that's only three words and I know where it comes from. I know what poem it's in, and I have an idea of what comes before it and what comes after it. So in terms of tunneling into my own work, the fact that I can recognize a poem and its subject and its history and its currents through three words—I think it's pretty interesting.

I suggest calling the moment when a poet looks at the output text of a poetry generator that processes his or her past oeuvre of nondigital poems and when he or she is struck by a specific phrase in this output text "the punctum of computer-generated poetry," to borrow a term that Barthes used in his analysis of photography. Barthes argues that a photograph can impact the viewer in two ways. The first, which Barthes called "studium," takes place when a photograph speaks to the viewer's general knowledge and culture and therefore causes him or her to "take a kind of general interest" (Barthes 1982:26). The photographer can stylize, that is, intend to produce, such an impact because he or she shares or is aware of the viewer's knowledge and culture.

The second kind of impact, which Barthes called "punctum," differs from the first in a number of ways. First, the punctum is the result of a detail in the photograph that the photographer did not intend to represent and that the viewer did not intentionally search for but that he or she finds to be "at once inevitable and delightful" (Barthes 1982:47): "it is not I who seek it out (as I invest the field of the studium with my sovereign consciousness), it is this element which rises from the scene, shoots out of it like an arrow, and pierces me" (26). Second, this detail is the result of a chance event, an accident: "A photograph's punctum is that accident which pricks me" (27); "from [the] *Spectator's* viewpoint, the detail is offered by chance and for nothing" (42). Third, the force of this impact results from a kind of resonance between the accidental detail and the viewer's personal circumstances, history, and makeup. It involves "the individual viewer's sheerly personal response" (Fried 2005:543) and is "an artifact of the encounter between the product of [an] event and one particular spectator or beholder" (546). Fourth, the photograph acquires a new and different value as a result of its punctum: "A 'detail' attracts me. I feel that its mere presence changes my reading, that I am looking at a new photograph, marked in my eye with a higher value" (Barthes 1982:42). Finally, "the effect it produces upon me is not to restore what has been abolished (by time, by distance) but to attest that what I see has indeed existed" (Fried 2005:555–556).

All of these features are present in some form in the previous vignette, with the crucial difference that Sarah is both the writer and the reader of the artwork being considered. First, Sarah did not search for the line "you never tongued above me" prior to reading the output text, nor did she intend for this line to appear in the output text when she created her poetry generator. Second, the appearance of this line and its specific location in the output text are, by definition, chance events. Sarah used her previous nondigital poems as raw material that the computer, based on the code she wrote, broke down into individual words and phrases that it then recombined randomly. Third, this line impacts Sarah because it resonates with her personal history—again, by definition, since the output text consists of bits and pieces of her past nondigital poetic oeuvre. Sarah likes this line not only because of its poetic value, but also because she knows that it appears somewhere in her own past work. It is unlikely that this line impressed the other workshop participants as much it did Sarah, because they do not share her personal history. Fourth, Sarah gains a new and different appreciation for the output text because it contains this specific line. Last, because this poetry generator uses her past work as raw material, this line's effect on Sarah has, indeed, a strong temporal dimension. Both at the beginning of the vignette, when she cannot recall where in her past work this line appears, and later, when she finally remembers this detail, the line's effect on her is, as Michael Fried described it, to bear witness that this concatenation of words did indeed exist.

The punctum of computer-generated poetry acquires its value against the backdrop of the fact that the aesthetics informing this kind of poetry aligns with Barthes's "studium." A photograph that is only defined by studium is a "unary photograph" that "emphatically transforms 'reality' without doubling it, without making it vacillate . . . ; no duality, no indirection, no disturbance" (Barthes 1982:41). To be sure, the mere fact that words appear in a double role in the computer-generated poems discussed and written at the workshop—that is, as both symbols that have meaning and as concrete sonorous and visual sign vehicles that, when joined together in a certain way, create textures that differ from while adding to the symbolic plane—makes it difficult to argue that such poems are not based on a strong dimension of duality and indirection and that they do not "double" reality and make it "vacillate." However, the powerful aesthetic ideal of creating textual patterns, which informs the kind of poetry generators that were the main focus of the workshop, nevertheless makes their output texts not entirely different from the "unary photographs" that Barthes describes as being "homogenous" and characterized by a "'unity' of composition . . . received (all at once), perceived. I glance through them, I don't recall them; no detail (in some corner) ever interrupts my reading" (Barthes 1982:41). These descriptions resonate with Jason's description of computer-generated poetry's aesthetic essence as a "texture of language" that is based on repetition and variation and as "something that's more like a river [that is] neither surprising nor suspenseful." When some workshop participants expressed their desire to experience textual specificity and particularity, it was against the backdrop of, and as a kind of reaction to, this general aesthetics and its instantiation in computer-generated poetry's textual abundance and excess, as well as its formal abstraction.

RHETORIC VERSUS POIESIS

In chapter 6 I argued that writing computer-generated poetry cannot easily follow the normative Romantic ideal of poetic inspiration, with its tropes of possession and expressive immediacy and spontaneity, because it is a process that involves delay, deliberation, and intentionality. I added that the practitioners of computer-generated poetry I worked with consequently argue that this art form can accommodate a critically informed and socially minded creative orientation. I gave the example of Sarah who, on the workshop's fourth day, explained to the participants that she was interested in exploring how digital algorithms might function as an infrastructure of social injustice, a point she demonstrated by discussing and modifying the code of *A House of Dust*. Sarah added in that exchange:

> So a program might replicate a certain kind of speech that dates to a certain time in which there was a lot of repression going on, on to the end of Jim Crow in the 1960s. And I mentioned Gwendolyn Brooks earlier, who is a very important poet to me.... She has a book called *Riot* [1969]. It engaged with the riots of the 1960s—important social issues in Chicago. I used Brooks's book as a source to use with the code of *A House of Dust*.... So this is my version of it. Same thing, random choice, but the [source] material is different. These are all words drawn from Brooks's poems [Sarah points at a list of words that she compiled from Brooks's poems, which appeared in the code that was projected on the screen]. Brooks was also interested in the same social issues such as modern architecture as an instrument of oppression, so I am very much interested in experimenting with intertextuality and using it to explore important social issues for me.

Sarah states that her purpose in experimenting with computer-generated poetry is strongly motivated by her desire to engage with social issues that have to do with different forms of social inequality and injustice. This desire informs her form of engagement with the code of *A House of Dust*, which contains words that today would be considered offensive. Sarah replaces this offensive language with words taken from the poetic work of an African American female poet, Gwendolyn Brooks, who, like Sarah, used her poetry as a platform to advance social causes.

The fact that writing computer-generated poetry is a process that involves delay, deliberation, and intentionality affords rather than determines Sarah's approach to computer-generated poetry as a form of critical practice. In other words, delay, deliberation, and intentionality are necessary but not sufficient conditions for her poetic stance. The example of Calvino, one of the key representatives of the Oulipo's second generation of writers—a generation that emphasized the importance of giving more space to semantics or meaning rather than only to syntax and to language as a concrete object—exemplifies this distinction. Calvino attempted to fashion literary constraints at the level of plot or narrative in a story he called "The Burning of the Abominable House." Fashioning these constraints was a process that required much delay, deliberation, and intentionality. These constraints include the following (Calvino 1998:145–146):

> 4 characters: A, B, C, D.
> 12 transitive, nonreflexive actions (see list below)
> All the possibilities are open: one of the 4 characters may (for example) rape the 3 others or be raped by the 3 others.
> One then begins to eliminate the impossible sequences. In order to do this, the 12 actions are divided into 4 classes, to wit:
> appropriation of will (to incite, to blackmail, to drug)
> appropriation of a secret (to spy upon, to brutally extort a confession from, to abuse the confidence of)
> sexual appropriation (to seduce, to buy sexual favors from, to rape)
> murder (to strangle, to stab in the back, to induce to commit suicide)

Calvino describes a number of "objective constraints" that relate to the "compatibility between relations" (Calvino 1998:146):

> For the actions of murder: If A strangles B, he no longer needs to stab him or to induce him to commit suicide. . . .
> For sexual actions: If A succeeds in winning the sexual favors of B through seduction, he need not resort to money or to rape for the same object. (ibid.)

These objective constraints are meant to prevent illogical plot lines. Without them, the algorithms would, for example, allow "Clem to buy sexual favours from Dani even after the latter has had her throat cut" (Duncan 2019:113). As becomes clear from this description, Calvino's sense of semantics and meaning is highly formalistic. Calvino is not interested so much in the specific meaning and semantic implications that might be produced by the constraints that he fashions as in making sure that the story has a logically coherent meaning at the level of plot or narrative. The fact that his story "replicates," to use Sarah's phrase, obnoxious "transitive, nonreflexive actions" such as raping, drugging, cutting someone's throat, strangling, and murdering for no apparent good reason is irrelevant to the overall point of his formalist-algorithmic literary exercise.[9]

Calvino's approach can be related to the fact that, as key members of the Oulipo such as Queneau argued, the Oulipo's literary endeavors are historically related to different forms of "mathematical entertainments" and "diversions" whose purpose is to "amuse" (Queneau 1998:52): "even at its most polemical, even at its most ferociously doctrinaire, the Oulipo's work over the past twenty-five years has consistently been animated by a most refreshing spirit of playfulness. The Oulipian text is quite explicitly offered as a game, as a system of ludic exchange between author and reader" (Motte 1998:20). It is not surprising, then, that the same ludic dimension is a strong feature of computer-generated poetry, too, as can be seen in the "humorous and ridiculous" effects of literary works that are heavily predicated on "digital (i.e., programmatic) foundations" (Funkhouser 2007:48; see also 52). The logical end of this ludic emphasis in the context of a form of poetry that prioritizes formal-logical constraints is "code poetry," or poetry written directly in a programming language and for which the code consequently need not be executed by a computer but can be read as a poem:

> Indeed, the impulse to put "poem" in quotes when discussing code-work "poems" comes from its constraints on human language, which result in a narrowness of vocabulary, rhetorical strategies, flexibility of language and the other pleasures that make poetry, in the traditional sense, poetry. With the constraints of the machine straightjacketing human language, readers of code poetry are often treated to techno in-jokes. . . . Haiku is the dominant form, not for its minimalist imagery, but because its short, disconnected lines can be fitted into a computer command, while the lines of code poetry themselves often exhibit the literary sophistication and nuance of greeting-card verse (and indeed, "Black Perl" [a well-known code poem] began life as more joke than poem). (Tomasula 2012:489)[10]

From Sarah's perspective, the algorithmic production of literature for the sole purpose of diversion or amusement can be a problematic formal exercise because it can easily lead to moral agnosticism and social irresponsibility. Similar kinds of concerns have been voiced in relation to other forms of digital art. For example, Nick Montfort has raised the following questions in relation to the field of interactive digital narrative: "Is it supposed to be pure entertainment or something beyond that" such as caring "about empowering the disenfranchised[?] . . . Is it about human nature and our cultural world, or a distraction from them? . . . Is it a pleasing but meaningless salve to soothe workday aches or a way to model a better society?" (2015:xi). Montfort suggests that a socially minded general orientation may be a stronger unifying factor for practitioners of digital art than the specific kind of digital art that each of those practitioners might choose to focus on: "The Twine [a specific form of interactive digital narrative] community . . . might decide that someone writing a Twine game that is, for instance, a humorous Choose-Your-Own-Adventure parody aimed only at laughs isn't really part of their movement, while a digital poet working with a different system could be" (2015:xiii). He also acknowledges that these two general orientations "can, of course, both exist, and there can be groups, even large and organized ones, working in both of these ways" (ibid.).[11] Montfort's perspective can be subsumed under a long tradition of critiquing the strongly conservative and apolitical tendencies that have characterized computer art from its very beginning (Taylor 2014:227).

Sarah emphasizes the importance of the social ends that the computer-generated poetry she writes can serve. This socially conscious orientation leads her to be mindful of the choices she makes in appropriating, modifying, and experimenting with existing code. She prefers to engage with computer-generated poetry because of "how much it foregrounds, or can foreground, rhetorical choice and intention," as she put it in the exchange I discussed in chapter 6. Significantly, of the five guest lecturers, she was the only one who stated outright that she has no problem with editing the output texts of her code if doing so might better serve the goals of her poetic practice. "I'm not wedded to the output," she told the participants. "It's not especially precious to me that this [her code] produce something that I would consider a golden poem." Sarah's practice contrasts with the strong emphasis expressed by the other facilitators on editing only a poetry generator's code rather than its output texts.

Sarah's framing of computer-generated poetry's purpose in terms of "foregrounding rhetorical choice and intention" is significant against the backdrop of the fact that at different points during the workshop facilitators contrasted rhetoric with poiesis and argued that because of its text-distributional aesthetics, computer-generated poetry lends itself more naturally to poiesis rather than to rhetoric. For example, on the workshop's second day, after Grant presented his poetry generator, a discussion followed about the meaning of its output text, after which Jason said: "One thing I wanted to mention is that this text has a point. You wanted to create something that is an interesting text. One question that we can ask is whether poetic works have to have a point. I think that rhetorical works have to say something. But does a poetic work particularly have to have a point?" In a conversation with Jason the following day, I asked him to clarify this distinction. He told me: "Rhetoric is about making or representing a specific point of view or position, versus poiesis, which is about opening things up, inquiring, asking questions, producing, making things. Of course computation can be more or less poetic, but computer-generated poetry is actually very good for creating textures and distribution-like things that raise questions and open things up for inquiry without making a clear point or reflecting a specific position."[12] Jason argues that computer-generated poetry lends itself more easily to generating interesting linguistic textures that do not necessarily support a specific position but can nevertheless "raise questions" and "open things up for inquiry" more generally. His argument aligns with critical perspectives that emphasize the predominance in new-media poetry of the "rejection of a single rhetorical authority and of linear causative organization" in tandem with "admission of aleatory organizations and relationships as more accurate representations of experience, and at least an effective illusion of the simultaneity of experience" (Funkhouser 2007:19).

Although in principle computer-generated poetry's text-distributional logic can also function as a vehicle for social commentary, throughout the workshop this function remained secondary to an emphasis on the production of abstract linguistic textures, rigorous logical-formal constraints and principles, and ludic effects. The performance-like event that took place in the evening of the workshop's last day, in which each participant presented a computer-generated poem that was the result of his or her participation in the workshop, exemplified this general orientation. The poems that most of the participants presented, as well as the evening in general, reflected the kind of powerful pull toward "mathematical entertainments" and "diversions" whose purpose is to "amuse," which the first generation of the Oulipo writers embraced as the rationale for their literary endeavors (Queneau 1998:52). To begin, Jason presented each of the participants by making different observations about their names, which he had prepared in advance and written on a piece of paper. Such observations included how some of the names can be anagrammed, the palindromes they contain, and the fact that some of the names consisted of letters that are symmetrical on the same or different axes. For example, when he presented my name, Jason said: "Our next speaker is Eitan. And if I were to tell you a narrative that was constructed of the substrings anagrammed from his name, it would be: 'I tie a tie at nite,' spelled N-I-T-E." Jason's mode of presenting the participants elicited laughter from the people who attended the event, including the workshop participants. It both demonstrates how the writer of computer-generated poetry observes the world through the prism of the ways in which it can be represented and manipulated in computational terms and resonates with the idea that "the Oulipian text is quite explicitly offered as a game, as a system of ludic exchange between author and reader," and that the spirit of such a text is the "spirit of playfulness" (Motte 1998:20).

My own presentation made me feel awkward because the computer-generated poem that I presented deviated from this ludic path. For my poem,

I took three condolence letters written by (or on behalf of) three former US presidents—Abraham Lincoln, George W. Bush, and Barack Obama—and sent to the families of three fallen soldiers. I divided each letter into lines of between one to seven words and wrote a program that takes the lines of each letter and rearranges them randomly. My purpose, as I explained during the presentation, was to explore the tension between the fact that, on the one hand, these letters are the product of a highly regimented, ritualized, and even clichéd public genre and, on the other, are meant to express heartfelt emotion in light of another person's deeply personal tragedy. By breaking the letters into lines and then rearranging them randomly, I hoped to denaturalize the phrases used in such letters—in some cases undermining their intended meaning altogether—and point to their conventionality.[13] As I read an output text of my poem, I felt that it cast a shadow over the lighthearted atmosphere generated by most of the poems that the participants read before (and after) me. Although I took comfort in the fact that during my reading and explanation about my poem's rationale, Sarah looked at me and nodded her head with what seemed to be an approval or encouragement, the feeling that my computer-generated poem was misaligned with most of the poems written by the rest of the participants continued to trouble me as I made my way back to my hotel after the evening ended and we all said our goodbyes. It troubled me not as an aspiring writer of computer-generated poetry, but as an anthropologist who had just realized that he had failed to embody one dimension of the poetic spirit of the people whose ways he had aimed to study, at least with respect to the limited context of the workshop's final evening. However, this failure contained a valuable lesson inasmuch as it highlighted some of this spirit's distinguishing features and tensions.

CONCLUSION

Neither Our Doom nor Our Salvation:
Open-Ended Digital Systems and Cultural Critique

The preceding chapters provide a comparative ethnographic basis on which it is possible to make a few concluding theoretical suggestions. These suggestions are best articulated in relation to the arguments made by scholars who have studied the design and use of the roboprocesses I briefly described in the introduction—namely digital systems to which different levels of decision-making are now delegated in such domains as criminal justice, banking, health care, education, agriculture, communication, and defense.

To begin, if roboprocesses are new forms of old bureaucratic techniques of control and management that consequently reflect state, institutional, and business interests indifferent to and even clashing with the interests of individuals (Gusterson 2019:4), the digital systems that are the focus of this book are the product of individuals' attempts to address specific predicaments that trouble them. My interlocutors were motivated to design and experiment with such digital systems because of specific concerns they had in relation to their creative practices, for which they hoped these systems could provide a solution. Consequently, they were able to recognize their own interests in the systems they designed and experimented with.

Another way of articulating this difference is that whereas roboprocesses are almost always imposed from above, the digital systems I discussed in this book emerge from below. This crucial difference has a number of implications. First, because roboprocesses are typically based on proprietary knowledge and provide the basis for economic profit or for pursuing state interests such as surveillance, their technological architecture and logic are frequently black-boxed (Besteman 2019:168–169; Gusterson 2019:11). Consequently, the consumers and citizens most affected by roboprocesses have no way of gaining insight into how they work. Even when they can, they are often unable to contest and change the outcomes of these processes (Eubanks 2018; Noble 2018; O'Neil 2006).

For example, Gusterson provides a vignette about his own experience with the roboprocesses that animate the banking system. After closing his bank account, a computerized system reopened it because of a remaining unpaid balance of a few cents stemming from a final interest payment. The unpaid balance gradually increased (without Gusterson's knowledge), because the bank charges customers with low balances a fee, until the resulting debt was finally sent to debt collectors. Observing the unsuccessful attempts made by a branch manager to address the problem, Gusterson concludes:

> Computerized processes . . . , while supposedly the embodiment of a rational system, . . . produced an absurd outcome that defied common sense. I, the customer, ostensibly served by the system, was trapped within it. The operators of the system, supposedly its masters, are disempowered, and it becomes hard to find anyone who has the authority to override the system's flaws. The algorithmic processes that underlie it take on a life of their own. . . . The common sense and situational logic of humans is displaced by and subordinated to the logic of automation and bureaucracy. (Gusterson 2019:2)

In contrast, my interlocutors have a higher-level knowledge of the architecture and logic of the digital systems that they interact with and that impact their creative practice because they are their designers. Consequently, even when their systems produce the kind of unintended consequences that roboprocesses are notorious for, my interlocutors have more agency in addressing them. They can tinker under the hood of their systems to change whatever in their architecture they think produces those unintended results. Consider David's realization that the genetic-algorithms framework he chose as the basis for his system produced unintended consequences that subverted his original goals. David built his system because he was annoyed by his fellow musicians' repetitive playing, among other reasons. However, his system produced improvisations that relied on successful phrases, which became repetitive because the genetic-algorithms framework produces solutions that gradually converge around an optimal one. His system thus reproduced an aspect of the problem for which it was supposed to provide a solution. In response, David developed mutation operators that ensured diversity in a way that defied the genetic-algorithms framework.

Furthermore, because my interlocutors interacted with systems of their own design, they could respond to these systems' unintended consequences with humor and even aesthetic interest, as James's research team did in relation to Syrus's sometimes unintelligible improvisations, a result of its technological architecture's not taking into account, or not being able to produce, what its human interlocutors considered to be meaningful dimensions of musical improvisation. Where "the common sense and situational logic of humans is displaced by and subordinated to the logic of automation and bureaucracy" in the case of roboprocesses (Gusterson 2019:2), my interlocutors subordinated Syrus's malfunctions and idiosyncrasies, as products of its logic of automation, to their common sense and situational logic, at times even willingly exposing those malfunctions and idiosyncrasies, as when Kim refused to perform repair work when he musically interacted with Syrus, or when James exclaimed "Motor music!" when Syrus's arms hovered above the marimba without actually playing it.

On the surface, the reason for these differences is that I am comparing apples with oranges, that is, two different and incomparable kinds of decision-making digital systems. My interlocutors could approach their systems' unintended consequences with humor and aesthetic interest because the implications of those consequences were not as momentous to their lives as the irrational and unintended consequences that a roboprocess such as the one that animates one's bank account might have. However, consider the fact that roboprocesses also include the kind of computerized technologies that "crunch the data that Facebook, Amazon, YouTube, and Google use to decide what you will be prompted to look at and in what order" (Gusterson 2019:6). These technologies are not unrelated to the technological architecture that animates some of my interlocutors' digital systems: in both cases the synthesis and mediation of different kinds of style is a key objective. They have been the focus of attempts made by search engines companies, social media companies, and a host of other entities to predict online users' individual preferences, tastes, and dislikes—in short, their styles—based on their online behavior, and then to provide and produce online content that could mirror and anticipate these styles (Wilf 2013d:737). Such digital consumer-centered production derives a large share of its profitability from its purported ability to tap into each consumer's distinct patterns or styles of behavior, especially when this behavior takes place online, because the online platform provides companies with vast data on consumers' past patterns of online behavior. At stake is these companies' attempt to replicate a technologically based author who can speak on behalf of the principal that is the consumer. When companies manage to achieve this task, consumers feel that the online content resonates with "who they are."

A growing number of scholars have argued that as a result of how companies use such digital technologies, individuals no longer receive the same online content. Rather, each consumer is sent content customized to mirror his or her style, a situation that leads to a "filter bubble," a stifling self-referentiality, narcissism, atomism, and fragmentation of the public sphere (Pariser 2011:112–113, 160). In addition, they have argued that it would be naive to think that commercial companies limit themselves to this task. Companies often strategically provide consumers with content that differs from rather than reflects their individual styles in order to modify those styles and make them more predictable and aligned with the companies' contents, products, and interests. Such technologies end up being performative rather than reflective of consumers' styles. This performative function amounts to the coproduction of users and technologies. Users and technologies react to each other and shape each other's specific style with a potentially asymptotic "fit" between them, whose goal is to make consumers' intentions and predilections more predictable and thus a better and more reliable source of profit (112).

These criticisms align with those I discussed in the previous chapters, which point to narcissism as a potential unintended result of interacting with a relational artifact designed to mirror the human user's preferences (as in David's case), and to the performative production of style under the guise of its reproduction (as in Syrus's case). Indeed, they draw from the same intellectual sources: "Roboprocesses . . . spur the zombification of interpersonal relationships. . . . Writing about efforts to design programmable robots whose algorithms can be individually tailored to make them totally compliant, undemanding, and responsive to their human companions, Sherry Turkle worries that the result will make us less human, less resilient, less tolerant, less flexible, less alive" (Besteman 2019:174).

In addition, critics have argued that a possible remedy for the filter bubble is to force companies to give each consumer control over the content he or she receives by, for example, crafting "an algorithm that prioritizes 'falsifiability,' that is, an algorithm that aims to disprove its idea of who you are" (Pariser 2011:233). Some of the suggestions for such falsifiability uncannily resemble the scene of "cocktailing of styles" with which I opened this book: "Google or Facebook could place a slider bar running from 'only stuff I like' to 'stuff other people like that I'll probably hate' at the top of search results and the News Feed, allowing users to set their own balance between tight personalization and a more diverse information flow" (235).

The fact that roboprocesses and the digital systems that are the focus of this book are related invites a reevaluation of the strong dystopic assessments of roboprocesses. As I argued in the previous chapters, designing and interacting with relational systems did not make my interlocutors "less human, less resilient, less tolerant, less flexible, less alive" (Besteman 2019:174). If anything, it encouraged them to be more resilient, more tolerant, and more flexible, because they had to cope with and respond to their systems' occasional malfunctions, interactional incompetence, or semantic unintelligibility. This experience, in turn, has made them more human and more alive because the challenges of digitally simulating creative agency brought into relief and made them reappreciate the complexity of their own human creativity.

Similarly, the clear-cut arguments that roboprocesses lead to "deskilling and disciplining" (Besteman 2019:174; Gusterson 2019:13–16) and that they tend "to amplify and naturalize inequalities and identities" (Besteman 2019:176) should be qualified, given that the design and use of digital systems in some contexts require one to learn new skills and affords the self-reflexive denaturalization of existing forms of inequalities and identities that hitherto have gone unnoticed. Consider the poets I worked with. Not only did my interlocutors have to learn new coding skills and use them to realize specific normative ideals according to a specific cultural logic (such as a text-distributional one), but some of them also took advantage of the fact that delay, intentionality, process, editing, and collaboration are fundamental dimensions of writing computer-generated poetry in order to make this art form a critical tool for challenging forms of social injustice in general, and forms of injustice that are the result of art's fetishized status in the modern West in particular.

In addition, their self-critical creative practice should give us pause regarding the argument that "the algorithmic self is also the regulated self" and that, in relation to "the algorithmic self," "self-management is outsourced to technology" and "choice becomes compliance" (Besteman 2019:178). My interlocutors in the field of computer-generated poetry took advantage of the option that digital computation gave them to "self-manage" their poetic selves by tracking their poetic tendencies and fixations, with the purpose of exploring new creative horizons and expressive avenues. In doing so, they used their technologies to denaturalize the modern Romantic notion of the self as an organic whole that one cannot escape. Indeed, against the backdrop of the argument that roboprocesses tend to "trap" customers "within" them (Gusterson 2019:2), the practitioners of computer-generated poetry I worked with used their digital systems to escape from their ossified poetic identities-as-destinies, which they experienced as a kind of prison. Against this complex backdrop, the vision shared by "Turkle . . . and many others [of] a future stripped of democratic participation and dissent, oriented only toward the goals of profit maximization for the few" (Besteman 2019:178; see also Gusterson 2019:23) reflects a partial view of the implications of decision-making digital systems on contemporary and future modes of sociality.[1]

If my interlocutors' experiences problematize such overwhelmingly pessimistic predictions about the implications of roboprocesses, they also problematize the optimistic alternative futures that the authors of those predictions imagine, such as the following: "It would make an interesting thought experiment to imagine a fusing of bureaucracy and computers that was human-centered. In this counterhistory the algorithms used to regulate social and economic decisions would be transparent and debatable, and citizens would have free access to and control over the data collected about them; and system designers would work with customers and citizens, not just with government and corporate elites, to create processes that were responsive to those caught up in them (Gusterson 2019:6)." My interlocutors did design digital systems that were human centered. They designed systems whose goal was to offer solutions to the different problems that troubled them. Similarly, because they designed or had control over the algorithms that animated their systems, they experienced those algorithms as relatively transparent and debatable, that is, subject to explicit commentary, critique, and change. Finally, my interlocutors were both system designers and the customers or consumers of their designed systems. In short, my interlocutors inhabited the kind of utopian counterhistory envisioned by the authors of the dystopian accounts of the roboprocessual present.

It is telling, then, that my interlocutors' experiences share some key aspects with the experiences generated by roboprocesses described in the aforementioned accounts. This similarity means that we should be wary of Manichean views of computer-mediated, algorithmic forms of sociality. For example, if roboprocesses are characterized by "their often relentless ability to presume and induce standardization and to fail when confronted with the nonstandard" (Gusterson 2019:20), both David and James unintentionally ended up reproducing different dimensions of the standardizing environment of academic jazz education that they had tried to escape by means of their systems. Similarly, if one problem with roboprocesses is that the computers that provide their infrastructure "offer a model of cognition that increasingly shapes our approach to the world" (6) and that roboprocesses "amplify" the "indifference" and "unaccountability" that are endemic to bureaucratic processes by automating and black-boxing them (4; see also 11), my interlocutors in the field of computer-generated poetry observed the world through a kind of filter that translates this world into phenomena and categories that are already amenable to computational representation and manipulation, and consequently many of them created poetic work they valued primarily for its algorithmic-formal-mathematical dimensions and rarefied abstract textual properties, rather than for the ways in which this work might relate to, and thereby provide a meaningful commentary on and critique of, society.

This comparison suggests that decision-making digital systems, like any phenomenon that is embedded in culture and society, have both potentialities and limitations whose realization and frustration are historically and culturally situated and determined. Even when consumer-citizens are these systems' designers, they can still end up being trapped in these systems' computational logic, the problematic repercussions of which can make them long for a pre- or nondigital computational world. And even when consumer-citizens are not these systems' designers, they can still find different kinds of pleasure and enrichment in the interaction with them. Neither our doom nor our salvation, such digital systems are and will remain saturated with the complexity of the social formations and cultural configurations that give rise to them and that provide the context for their use and for the evaluation of their results and implications. If, as scholars, we do not open ourselves up to these systems' multilevel open-endedness, we risk embodying and reproducing the deterministic and rigid cultural logic that we intend to critique.

Notes

Introduction

1. All names and locations in this book have been changed to maintain anonymity of research subjects.

2. *Yardbird Suite* is a jazz tune written by saxophonist Charlie Parker (1920–1955). The "head" of a jazz tune is its melody. Typically, musicians begin improvising on a tune after playing its head.

3. "Trading fours" is a practice in which different players improvise on a tune a few measures each, one after the other, and in response to each other's playing.

4. "Miles" is the jazz trumpeter Miles Davis (1926–1991). "Coltrane" is the jazz saxophonist John Coltrane (1926–1967).

5. Jason refers to Herbert Simon's classic 1969 book *The Sciences of the Artificial* (2019: 51–53).

6. At the time of my fieldwork, the style imitation focused mainly on pitch and duration values.

7. This book represents the continuation of my ongoing research focus on the different dimensions of this co-constitution as they find expression in different institutional sites in the modern West, such as academic art programs (Wilf 2014a) and the development and implementation of routinized strategies of business innovation (Wilf 2019).

8. See https://www.fi.edu/history-resources/automaton, accessed July 12, 2020.

9. See https://www.theparisreview.org/blog/2016/03/18/automaton/, accessed July 12, 2020.

10. Kenny Dorham (1924–1972) was a legendary jazz trumpet player.

11. *Billie's Bounce* is a well-known jazz standard.

12. Note, however, that under certain conditions, students who engage in such a practice can experience a ritual transformation in which they feel that they are the authors rather than merely the animators of the solos they memorized (Wilf 2012).

13. Freddie Hubbard (1938–2008) was another legendary jazz trumpet player.

14. This is not to deny that, to be successful, imitation requires active engagement and the making of many creative decisions on the spur of the moment, as well as that the distinction between a finite text and a style depends on specific contexts of interpretation, normative ideals, scales of perception, how these texts are used, and to what ends (Wilf 2012:39–40).

15. The history of jazz has been riddled with examples of and concerns about players trying to inhabit the improvisation styles of influential musicians. For the case of Charlie Parker as an object of widespread imitation, see Shim (2007:42, 75).

16. Bourdieu argued that "the schemes of thought and expression [the individual] has acquired are the basis for the intentionless invention of *regulated improvisation*" (Bourdieu 1977:79, emphasis added; see pp. 11, 17, and 54 for this notion of habitus as "regulated improvisation"). His use of the notion of improvisation is serendipitous in the context of one of the art forms that I focus on in this book, namely jazz.

17. Throughout the twentieth century, a number of foundational figures in anthropology appropriated the notion of artistic style to theorize culture as a relatively coherent and durable set or configuration of generative dispositions and behaviors, which individuals acquire through prolonged periods of socialization and are conditioned to enact in a quasi-automatic and consistent manner in different situations because these dispositions are anchored in the unconscious, in the body, or in all-encompassing symbols. In addition to Bourdieu, these figures include Gregory Bateson, Ruth Benedict, Franz Boas, Clifford Geertz, Alfred Gell, Alfred Kroeber, and Edward Sapir (Wilf 2013d).

18. The movie *My Fair Lady* (Cukor 1964), which describes Henry Higgins's attempt to help Eliza Doolittle acquire a new habitus, provides a number of excellent dramatized depictions of such a pretense. In the early stages of Eliza's "transformation," she, like Henry, feels what it means to pretend to have a certain habitus by memorizing a few of its fixed expressions, such as specific phrases, in advance of participating in social interactions with representatives of the higher class. Once she runs through her limited reservoir of phrases, she has no choice but to betray her "real" social origins because she lacks the generative principle that is the habitus of members of the upper class and is consequently ridiculed by her snobbish interlocutors.

19. See Wilf (2010:568–570) for the presence of these modern normative ideals in US academic jazz education.

20. For one example out of many, see http://www.vanityfair.com/culture/2012/10/wolfgang-beltracchi-helene-art-scam, accessed July 12, 2020.

21. See Wilf (2014a:6–7) and (2014b:401) for an analysis of this antinomy as it is featured in the work of foundational thinkers such as Max Weber, Franz Boas, and Victor Turner, and in the work of a number of contemporary anthropologists.

22. See the example of "Life Design," a self-help method meant to help people "innovate" their lives by means of design methods, and in which the same characteristics find expression (Wilf 2019:147–176).

23. Studies of computational creativity that accept Boden's typology as is—that is to say, uncritically—abound. For a recent example, see du Sautoy (2019), especially pp. 8–15.

24. Bown's call for the field of computational creativity to shift away from its current predominantly abstract and formal study of creative systems in favor of "looking at interactions between systems and humans using a richer cultural model of creativity, and the application of empirically better-grounded methodological tools that view artificial creative systems as situated in cultural contexts," is an exception (2014:112). Moreover, as one study that has taken up this call demonstrates, even a culturally situated approach to the study of creative systems can continue to reflect the field's overall focus on formalized and relatively abstract propositions (Kantosalo, Toivanen, and Toivonen 2015:282).

25. For a review of other music-improvising interactive digital systems, see Young and Blackwell (2016).

Chapter One

1. Clifford Brown (1930–1956) was a legendary jazz trumpet player.
2. I discuss Weizenbaum's system in greater detail in chap. 4.
3. The fact that jazz improvisation in particular, and improvisation more generally, depend on the improviser's possession of a preexisting stock of building blocks with which he or she can build a convincing narrative or "story" (Monson 1996) qualifies Lev Manovich's argument that "database and narratives are natural enemies" (Manovich 2001:225), as well as his description of "traditional cultures," as characterized by "well-defined narratives" and "little 'stand-alone' information" (217). It is precisely the existence of such quasi-"stand-alone information" in jazz that has driven many computer scientists to try to simulate it with computers.
4. Hayles (1999:57) provides an analysis of metaphorical slippage as a key mechanism that has helped to naturalize the popular equation between brains and computers—the view that a brain is a kind of computer, and that a computer is a kind of brain (see Helmreich 1998b:102–105 for another example of metaphorical slippage). Such slippage typically depends on the oversimplification of different domains so that they can be discursively represented as commensurable and analogical to one another (Hayles 1999:62). Another example of metaphorical slippage in the field of computational creativity in jazz can be found in François Pachet's discussion of his jazz-improvising system Virtuoso (2012). As I will discuss in further detail later in this and in the next chapter, Pachet first provides a rather reductive definition of bebop improvisation and virtuosity, which takes into account primarily melodic and harmonic dimensions, leaving out many other dimensions of virtuosity in this specific style, such as interactional dimensions. This oversimplification of bebop improvisation and its reduction to two of the musical dimensions that are the most amenable to digital computation then allows him to argue that "in some sense, jazz improvisation is a special form of computing" (2012:119).
5. The schema theorem, proposed by John Holland (1992 [1975]), formulates the computational basis for the convergence of solutions around an optimal one in a genetic-algorithms computational framework.
6. For an analysis of the hitherto unsolved problem of how to evaluate the creativity of computational systems, see Jordanous (2019).
7. See Ventura (2019) for an attempt to design an image generator whose architecture "incorporates neural networks that have been trained to assess the aesthetic quality, the artistic style, and the semantic content of images," with the purpose of allowing the system "to evaluate and understand its own artefacts" (55). Of course, the training process of a neural net might introduce additional complexity, depending on how the net is trained (Gervas 2019:291). Such attempts notwithstanding, evaluation of creativity in computational systems (both by the systems' designers, let alone by the systems themselves) has remained "a notoriously difficult question that the field [of computational creativity] as a whole has yet to satisfactorily resolve in the general case" (Ventura 2019:61; see also Cohen, Nake, Brown, Brown, Galanter, McCormack, and d'Inverno 2012; see Ritchie 2019:159–160 for the presence of this problem in relation to systems that operate in different domains). This perennial difficulty has led some researchers to argue that "computational creativity may never be satisfactorily demonstrated, only controverted" and hence that "just as scientific theories are constructed to resist falsification, system builders must construct systems that resist demonstrations against their creative potential" (Ventura 2019:65; see also Ritchie 2019:171).

8. The many kinds of statistical data that David amassed on the phrase and measure populations and that he liked to consult after training sessions also demonstrate the omniscience that many programmers enjoy having in relation to the microworlds they design.

9. In so doing, David decided to leave out not only internal evaluation, or evaluation that "is part of the model with which the system operates, guiding the computations and having an impact on which artefacts (or behaviours) are eventually displayed to the outside world," but external evaluation too, that is, testing the system's output/behavior against "pre-existing ideas, either about artefacts in a particular domain, or about strict creativity," which "originate outside of the system itself" (Ritchie 2019:166). David feels that he doesn't need to perform ongoing external evaluation of his system's performance because, by definition, the improvisations it generates will always mirror his preexisting ideas about what a good improvisation is.

10. See Hayles (2009:42) and Taylor (2014:170–171) for artists' fascination with the mix of control and lack of control that they can experience by programming nonlinear processes and by applying the discourse of complexity in the domain of computer-based art.

11. In the field of computational creativity more generally, systems such as David's (and, as we shall see, James's), which lack the ability to reflect on and evaluate their own creations, are somewhat pejoratively described as "mere generation" (Ritchie 2019:165–166).

12. The paradoxical nature of David's attempt to downplay the role of technology by means of technology is a common feature of what elsewhere I called "the mediation of immediation" (Wilf 2019:127–128). Indeed, his story about the Computer History Museum bears a striking resemblance to another example of the mediation of immediation that I analyzed in the field of business-innovation consulting services and that also involves an old Macintosh computer enclosed in a glass case, as if it were a specimen of an extinct species (Wilf 2019:130–131).

Chapter Two

1. See https://www.youtube.com/watch?v=lLULRlmXkKo, accessed August 23, 2020.

2. To be sure, the person who narrates the video says at one point, "Mrs. Walter's pet is Elmer, Elsie's brother," but the video is overall suffused by a kind of family-based imagery.

3. Lucy Suchman provides a critical analysis of this trope in relation to video demonstrations produced by Boston Dynamics, in which the company features one of its navigational robots, BigDog, which is a much more complex robot that is also modeled after an animal (Suchman 2012). This robot was designed to be used in warfare to help soldiers carry supplies and perhaps weaponry in challenging terrain. The video demonstrates BigDog's advanced mobility capabilities in different settings. Crucially, it also shows a person kicking BigDog, which immediately corrects its posture. Suchman comments on "the slavish subservience that the robots themselves materialize in concert with their human masters, exemplified in the act of kicking the robot that seems to be an obligatory element of every demonstration video, so that we can watch it stagger and right itself again" (ibid.). For Boston Dynamics' video, see https://www.youtube.com/watch?time_continue=41&v=cNZPRsrwumQ&feature=emb_title, accessed August 25, 2020.

4. The analysis of such malfunctions is the focus of chapter 4.

5. For an attempt to integrate the feature of playing outside the harmonic changes into the improvisations produced by a computerized system, see Pachet (2012:136–138).

6. For a discussion of these forms of contingency in the context of games as "domains of contrived contingency," see Malaby (2007).

7. David's evaluation is eerily similar to Pachet's evaluation of his jazz-improvising system

Virtuoso: "Considering melodic virtuosity as a specific mode, we claim that these automatically generated choruses are the first ones to be produced at a professional level, i.e. that only a limited set of humans, if any, can produce" (Pachet 2012:143). The similarity between these evaluations is important not so much because it qualifies Pachet's argument (since David's system is older than Pachet's), but rather because it demonstrates how the radical simplification of human creativity in general, and jazz improvisation in particular, to a very few dimensions such as pitch and duration as a condition of possibility for jazz improvisation's computerized simulation then leads the designers of such simulations to claim that their computerized systems can produce output that is superior to what humans can produce. I will return to this point later in this chapter, as well as in the next two, when I discuss the counterexample of an attempt to design a much more complicated jazz-improvising system that attempts to integrate some of the context-sensitive dimensions that are so essential to virtuosic jazz improvisation.

8. Two well-known and mythologized examples involve the saxophonists Charlie Parker and Sonny Rollins (Petrusich 2017).

9. The kind of tight interaction epitomized by the trading-fours format suggests that when this format provides the basis for a human-machine distributed form of creativity, the dichotomy of "computer-supported human creativity" and "human-supported computer creativity" (Gatti, Özbal, Guerini, Stock, and Strapparava 2019:239) becomes inadequate because both forms of creativity become co-constitutive.

10. I discuss the notion of repair-work in detail in chapter 4.

11. For an example of what such a simple tone generator sounds like in the service of a jazz-improvising computerized system and how it impacts the aesthetic quality of the improvisation, see Pachet's Virtuoso at https://www.youtube.com/watch?v=iGvWVy7dglk, accessed August 31, 2022.

12. *Giant Steps* is an up-tempo and harmonically challenging tune written and made famous by Coltrane in his 1960 record of the same name.

13. Although David is aware of his system's limitations, designers often downplay their systems' limitations that have to do with those dimensions of jazz improvisation that are context dependent and not easily amenable to computation. Such downplaying can lead to a misunderstanding or misrepresentation of what jazz improvisation consists of. For example, although virtuosity in jazz improvisation (including in bebop) significantly depends on the ability to display competence in the different context-based dimensions that I discussed in this and in the previous chapter, Pachet equates it mainly with the speedy or fast execution of ideas rather than, for example, with "slow improvisation," which he considers to be a different mode, perhaps because, as he acknowledges, it "is a most challenging mode for cognitive science and musicology, as it involves dimensions other than melody and harmony, such as timbre and expressivity which are notoriously harder to model" (2012:143). To reiterate a point I made earlier, by oversimplifying how virtuosity is understood in the cultural order of bebop improvisation and by limiting it to computable dimensions, Pachet can claim that his system displays virtuosity in jazz improvisation. Not surprisingly, one of the videos that demonstrates Pachet's system also features the system's improvisation on Coltrane's *Giant Steps*. For a link to the video, see n. 11.

14. Note that students' inability to respond to the playing of their bandmates in real time is also the result of their preference for practicing with a recorded rhythm section that plays over the harmonic changes of standard tunes instead of with a live rhythm section. Such recorded rhythm sections are similar to the nonreactive rhythm section that accompanies the improvisations generated by David's system.

15. It is not surprising, then, that other computerized jazz-improvising systems whose conceptual architecture is predominantly informed by chord-scale theory as codified in massproduced method books also display these forms of context insensitivity (e.g., Pachet 2012:130).

16. Of course, digital improvising systems need not be as heavily informed by this epistemological logic as David's is. In the same way that contemporary academic jazz education has acquired its present form for historically specific reasons that include specific racial politics, the structure of public funding for the arts, and the politics of literacy and accreditation in the United States (Wilf 2014a:25–52), David designed a system that excludes a number of dimensions that are part and parcel of the aesthetics of jazz because of his own contingent rather than necessary predilections and preferences. Other predilections and preferences would have resulted in a different system (which, of course, might have resulted in different unintended consequences). For example, George Lewis intentionally designed Voyager, "a nonhierarchical, interactive musical environment" (Lewis 2000:33), to display "simultaneous multiplicities of available timbres, microtonal pitchsets, rhythms, transposition levels and other elements . . . [that are] emblematic of an aesthetic of multidominance" (36), in large part because he was inspired by African and African American music-making traditions, as well as by his membership in the Association for the Advancement of Creative Musicians.

Chapter Three

1. See Katherine Hayles's analysis of an early articulation made by Norbert Wiener (1954 [1950]), one of the pioneers of cybernetics, of the differences between rigid machines (associated with the Industrial Revolution) that threaten to co-opt human beings' flexibility and creativity (as beautifully demonstrated in popular cultural texts such as Charlie Chaplin's movie *Modern Times*), and cybernetic—that is, flexible—machines that can help human beings to cultivate their creativity (Hayles 1999:105).

2. Chick Corea (1941–2021) was a legendary jazz pianist.

3. See, however, one of Pachet's later systems, Flow Machine, designed "to play in one style while taking constraints from another. . . . In one instance, [Pachet] took Charlie Parker's style of the blues and combined it with the constraints offered by the serial world of Pierre Boulez" (du Sautoy 2019:208).

4. For additional images of Syrus's technological infrastructure, see Wilf (2013d).

5. For other examples of style mixing that involve jazz, see Boden (2003:312) and Cope (2005). Compare this mash-up of styles and its conditions of possibility and implications with new developments in the field of bioinformatics, which allow "lateral translations" of genetic material between, and "horizontal leaps across," different branches of "the genetic classification trees of species that first began to emerge during the 19th century" (Mackenzie 2003:321). These developments, which are enabled to a large extent by the algorithmic processing of genetic material as digital information, are creating new kinship and biopolitical imaginaries (Helmreich 2003).

6. The logical end of a high order sequence is a sequence of notes that matches the entire sequence of notes played in a given solo; it is, in fact, a copy of the solo.

7. See Pachet (2012:127–129) for the use of Markov chains in jazz improvisation generators, and Gervas (2019:292) for similar considerations in poetry generators that are based on n-gram models of language.

8. Randy had other things to say about the ways in which the well-crafted demos that featured Syrus might give viewers the wrong impression about its capabilities: "People go to the

internet and they see these videos, where some of the musical pieces are sort of pre-composed. These pieces are cool, but maybe people go and see these videos and then say: 'OK, Syrus can do this on the fly. I can play my consonant Western tonal piece and then Syrus will play like this on the fly,' which is not true. They need to know. I am not saying that that is worse than what Syrus does do on the fly. I honestly don't think it is. But they need to know." Such well-crafted demos, which obscure as much as they reveal about Syrus's true capabilities, bring to mind the common tendency of computational creativity researchers to omit the "systematic evaluation" of their artificially creative systems "in favor of the presentation of hand-picked star examples of system output as means of system validation" (Gervas 2009:61; see also Suchman 2012).

9. This orientation is essentially based on avoiding what a computational creativity scholar has described as the need "to find if the model can accurately describe known exemplars within the domain"—in other words, whether "there is a way to run the program so that the output item is a known artefact" (Ritchie 2019:174), such as a specific improvisation produced by a specific past jazz master.

Chapter Four

1. Garfinkel's "groundwork for a sociological approach to information ... has been essentially lost to scholarship since it was first written" (Rawls 2008:7).

2. According to Garfinkel, for signals to have meaning they must be idealized and thus become signs. Technically speaking, then, the students are trying to separate idealized signals from noise.

3. A striking example of noise's potential to constitute the foundation of an aesthetic world is the Japanese noise-music scene. In his analysis of it, David Novak (2013) discusses the role played by cybernetics in the constitution of the anthropological and sociological imagination (although he does not discuss Garfinkel), as well as the potential of positive feedback loops, amply used in the Japanese noise-music scene, to complexify anthropology's and sociology's emphasis on the use of negative feedback mechanisms to maintain social order (Novak 2013:139–168; see also Larkin 2008:218).

4. Although everyday semiotic processes consist of signs such as indices and icons whose materiality is inherent to their ability to signify, under Enlightenment-era semiotic ideologies the crucial role played by such signs has been suppressed, even as they continued to provide the foundation for virtually every kind of meaning-making process, including the kind of scientific reasoning celebrated by the architects of the Enlightenment (Peirce 1992). Icons and indices can thus signify by virtue of their materiality without this materiality's becoming an object of awareness. Furthermore, although the materiality of icons and indices is crucial to the ways in which they signify, such materiality often does not reflexively point to itself (as it does in uncanny phenomena), but rather points to other objects.

5. My anthropological definition of the uncanny entails that the uncanny will not be a salient phenomenon in cultural contexts in which the materiality of semiotic forms is always already salient (see, e.g., Robertson 2018).

6. See Irani (2015) for a similar argument in relation to artificial intelligence, and Schüll (2012:174) for a similar argument in relation to the design of gambling machines.

7. In attempting to advance this goal, the research team drew eclectically from different sources, including current neuroscientific research on joint action and Konstantin Stanislavski's method of acting, but not from Garfinkel's theories.

8. Team members were not immune to the illusion of authentic self-expression that Syrus's embodied form could generate. As Randy told me a few months after he had left the lab: "I had separation anxiety after leaving the lab. Because in my three years there, probably the one person [sic] I spent the most time with was Syrus. And I think that as soon as we built the head that looks at you and kind of breathes and it got sound and even the motors going on and off and up and down, it definitely has a very—it's not just a machine anymore."

9. Of course, such techniques of estrangement need not necessarily function as forms of critique. They can also end up functioning as techniques of obfuscation (Kockelman 2016:27–53; Lemon 2018:186). This seems to be the case with the "behind the scenes" footage that one can now find at the end of many action movies, and that represents the genred commodification of the unmasking of only some and not necessarily the most crucial conditions of production.

10. These practices might be a form of "bothunting," that is, purposefully engaging in humanlike behavior with performing objects in order to expose their lack of humanity. These are "in effect Turing tests asking 'are you human?'" (Manning 2018a; see also Manning 2018b:69), here performed for humorous rather than for detection purposes.

11. As Fred Turner (2006) shows, however, these theoretical frameworks would later provide the conceptual and technical foundations for different social-cultural projects and movements in the United States, which focused on the advancement of personal liberation and utopian communal ideals, in opposition to cold-war conservatism.

Chapter Five

1. For code as a kind of underlying unconscious of the Information Age, see Hayles (2006b).

2. Inasmuch as Google Translate has come to represent a reliable form of "detecting" a language (as opposed to translating it), this vignette thus turned out to be doubly ironic. Google Translate—a machine-based translation device—was fooled into giving a nonsensical response to an equally nonsensical query produced by another machine-based device: Shiv's code.

3. Throughout the workshop, the facilitators brought up the Oulipo time and again as a literary frame of reference for computer-generated poetry. I will discuss the continuities and discontinuities between computer-generated poetry and the Oulipo in the following chapters.

4. This translated selection is taken from: http://stuttgarter-schule.de/lutz_schule_en.htm, where it is also possible to find Lutz's explanation about his poem.

5. Jason's argument also applies to text generators that do not produce a finite or "exhaustive" text.

6. There was an important exception to this stance, which I will discuss in chapter 8.

7. Narrative has been more convincingly present in forms of digital art that incorporate the user's input. In such contexts, the narrative is constructed by means of, and changes according to, such an input in real time (Hayles and Montfort 2012; Koenitz, Ferri, Haahr, Sezen, and Sezen 2015; Montfort 2007).

8. The aesthetics of text distribution, too, exemplifies the logic of the database that underlines many new media objects in the computer age. Such objects "do not tell stories; they do not have a beginning or end; in fact, they do not have any development, thematically, formally, or otherwise that would organize their elements into a sequence. Instead, they are collections of individual items, with every item possessing the same significance as any other" (Manovich 2001:218).

9. See Gendolla (2009:171) for a sample output text of *Love Letters*, Wardrip-Fruin (2011)

for an extensive analysis of Strachey's poem and of his pioneering role in the history of digital literature, and Funkhouser (2007:60–61) for a description of *A House of Dust*.

10. Both Sandra's reservations about the repetitive nature of many computer-generated poems and Jason's defense of it problematize cultural critiques of repetition in other art forms such as minimal music. According to one such critique, "The single-minded focus on repetition and process that has come to define what we think of as 'minimal music' can be interpreted as both the sonic analogue and, at times, a sonorous constituent of a characteristic repetitive experience of self in mass-media consumer society" (Fink 2005:3–4). Sandra's reservations, which, as I will argue in the next chapters, were shared by some of the other workshop participants, reveals a critical self-reflexive stance toward repetition, which suggests that repetition is far from becoming the contemporary self's key constitutive logic to which the individual unreflectively succumbs. Jason's defense of repetition, e.g., his explanation that repetition can clarify the essence of repetitive phenomena such as the flow of a river for which narrative-based literary forms are inadequate, suggests that some repetitive cultural forms that are intimately related to minimalist music may be much more than manifestations of a mass-media consumer society, and perhaps have nothing to do with it.

11. I will elaborate on this point in chapter 8. For works that involve the editing of output texts, see Funkhouser (2007:68).

12. See http://theprostheticimagination.blogspot.com/, accessed August 31, 2022.

13. As the previous discussion about observing the flow of a river makes clear, however, these quantity-based entities—the resulting linguistic textures that are based on repetition and variation—can be (and often are) experienced and evaluated as qualities.

Chapter Six

1. For studies of American liberal subjectivity in relation to computing-based communities of practice, see Coleman and Golub (2008); Helmreich (1998b).

2. Indeed, inasmuch as American anthropology is also strongly informed by American liberal subjectivity, some of the contradictions that characterize it might not be entirely unrelated to those I trace in this chapter. To give just one example, this academic discipline currently celebrates the advancement of social justice and inclusivity as one of its key goals, yet it prioritizes high theory, which is anything but inclusive, as the means to pursue it.

3. See Novak (2013:139–168) for a discussion of musicians' lack of control as a normative ideal in the Japanese noise-music scene.

4. As I will argue in chapter 7, my interlocutors' rejection of such Romantic normative ideals of creativity, especially inspiration, has some of its roots in the French literary group Oulipo.

5. Of course, digital art can combine these two forms of randomness—the one "already existing in the world," the other generated by the computer (Hayles 2009:39–40).

6. This is the case despite the relatively simple computational architecture that my interlocutors use and that led Jason, in the second of the two vignettes with which I opened this book, to emphasize to the participants that a poetry generator is a simple mechanism akin to an ant and that the complexity that can be discerned in a generator's output texts lies with the reader in the same way that the complexity of an ant's path lies in the beach on which it walks.

7. Such descriptions suggest that whereas critics have tended to conceptualize the use of computer-based chance procedures in art as a form of authorial abnegation or nonintention and as a reaction to Romantic notions of creativity, Romantic theories of creativity often iconized authorial abnegation or nonintention.

8. The strategy of redeeming nonsensical textual outputs by means of their sonorous linguistic qualities would become a recurring feature in the subsequent digital transformations of the modernist aesthetics that informed the Oulipo. Thus, in relation to some of the output texts of his computer-assisted literary works, John Cage writes that they "do not make ordinary sense. They make nonsense. . . . If nonsense is found intolerable, think of my work as music" (quoted in Funkhouser 2007:66).

9. The participants' reservations point to the limited scope of the following kind of analysis of a specific text generator: "Yet out of such nonsense we are able to detect poetic logic. Deeper meaning can be evoked when odd statements [are] encountered" (Funkhouser 2009:70). This statement clarifies the literary scholar's own understanding of the poem rather than how and to what extent members of a community of practice might be able to perceive "deep meaning" in the products of such a text generator in the context of real-time situated action.

10. Another question was, "How do you fashion narrative arcs when it is also at random?" This question takes us back to the kind of reservation expressed by Leandra.

11. The symbolically mediated and hence implicit form of socioeconomic exclusion that found expression in the rarefied aesthetics celebrated during the workshop might have already been heralded by the more explicit form of socioeconomic exclusion that found expression in the workshop's relatively high participation cost ($1000) and location in New York City—in other words, even before the workshop began—although a few subsidies were made available to some workshop participants.

Chapter Seven

1. Recall that Jason responded to my queries about how he comes up with ideas for new computer-generated poems and whether he has an identifiable style by invoking Georges Perec, one of the Oulipo's members. In a subsequent session during the workshop, which I will discuss in chapter 8, Jorge, the guest presenter, also brought up the Oulipo in relation to the kind of computer-generated poetry that was the focus of the workshop. The influence of the Oulipo on computer-generated poetry is a generally acknowledged fact in critical accounts of this form of digital poetry (Funkhouser 2007:33–34). As Jason's comments about Duchamp's association with the Dada movement suggest, computer-generated poetry might be informed by modernist aesthetic currents that date back to much earlier than the Oulipo. For the influence of Dadaism on computer-generated poetry, see Funkhouser (2007:32–33, 79).

2. In 1981 two members of the Oulipo, Paul Braffort and Jacques Roubaud, proposed creating "a new group dedicated to specifically computer-oriented Oulipian research" (Mathews and Brotchie 1998:46). The group was named Alamo, an acronym that in French stands for "workshop for literature assisted by mathematics and computers." See also Gervas (2019:281–282).

3. This principle points to an inconsistency in the Oulipo's overall philosophy. On the one hand, the notion of Ouvroir or "workshop," which was the result of lengthy deliberations among the group's founders, implies "a certain modesty in the group's aims, an anti-theoretical self-image which . . . sets them apart from the structuralists" (Duncan 2019:13). On the other hand, their disdain for "applied Oulipo" (Motte 1998:12)—that is, for exploring the implications of their self-fashioned constraints—betrays a strong theoretical snobbism.

4. An identical form of fetishization of the high number of potential texts that can be generated by means of a specific self-fashioned constraint can be found in an essay written by another member of the Oulipo, Italo Calvino, titled, "Prose and Anticombinatorics": "The possible

solutions, in consequence, are twelve to the twelfth power; that is, one must choose among solutions whose number is in the neighborhood of eight thousand eight hundred seventy-four billion two hundred ninety-six million six hundred sixty-two thousand two-hundred fifty-six" (Calvino 1998:144).

5. See Wilf (2019:90–94) for an analysis of forms of business-innovation strategies that combine the first and third meanings of potentiality as a result of their embeddedness in a capitalist cultural order.

6. Note, however, that the writer's absolute rational control and formalist prowess continue to take center stage even in the context of this conceptual framework, whose purpose is to open up a space of poetic freedom for the writer. Consider this description of the clinamen given by one of the Oulipo's members, Jacques Roubaud: "A clinamen is an intentional violation of constraint for aesthetic purposes: a proper clinamen therefore presupposes the existence of an additional solution that respects the constraint and that has been deliberately rejected—but not because the writer is incapable of finding it" (quoted in Duncan 2019:118). Whatever aesthetic freedom the clinamen is supposed to grant the writer in relation to a constraint, it is important for Roubaud to emphasize that this aesthetic freedom presupposes the constraint, the solution that this constraint imposes and that the clinamen replaces, and, most important, the writer's awareness of the constraint and the solution. This extremely rigid notion of aesthetic freedom finds expression in practice, too, not only in theory. The attempts made by the Oulipo's second generation of writers to introduce poetic freedom were highly formalized in that the kinds of deviation they experimented with in relation to their self-fashioned constraints were based on clear and rigid rules (Motte 1986:273–276). A critic who has studied George Perec's celebrated text *La Vie mode d'emploi*, described the following example of the clinamen in this text: "First, while the formal organization of the novel provides lists of forty-two constitutive elements to be integrated into each of the chapters, . . . two components of each list . . . function programmatically throughout the novel as clinamens; the former imposes the transformation of an element in each list, the latter the suppression of an element. . . . By this strategy, 'the dysfunction of the system is itself systematized'" (275). Note, too, that even members who belonged to the group's first generation allowed themselves to deviate from the rigid procedures they had devised in order to achieve more aesthetically pleasing results, although they did not explicitly acknowledge that they had so deviated (Duncan 2019:118).

7. For the case of Georges Perec, see Grimstad (2019). For an additional example of the reification of the poet as a quasi-natural genius who is a creator of new worlds in contemporary digital poetic contexts that are supposed to negate Romantic normative ideals of creative genius, see Wilf (2020:15).

8. For an analysis of contemporary manifestations of these principles in creative-writing workshops, see Wilf (2011:473–474).

9. A similar kind of argument has been made in relation to early examples of "computer poems": "Chance is not abolished by the computer's randomizing power but is re-created in different terms. The poet-programmer finds this power a tool to create a new set of dice, multifaceted and marked with elements of his own choosing" (Bailey 1973).

Chapter Eight

1. At one point during the workshop, Jason offered another response: "Another thing you can say is that poems are what poets are writing. Poems are what is included in poetry magazines. That's not a silly definition. It takes society into account. It takes context and culture into

account." Jason's response draws from a sociological—which is to say, institutional—perspective on art, in the context of which whatever is written by the people whom society designates as poets and is subsequently published in venues that society calls "poetry magazines" is poetry. This approach aligns with the sociological notion of "art worlds" (Becker 1982).

2. Such statements can be subsumed under "the prolonged attempt to submit art to the powers of mathematics" (Taylor 2014:18), an attempt whose history is centuries old.

3. Jorge refers to Jason's discussion of Simon's argument about the basis of the complexity of an ant's convoluted path on the beach, a discussion with which I opened this book. I return to this vignette later in the chapter.

4. Note that "Barthes was even touted [by the Oulipo] as a potential invitee in the 1970" (Duncan 2019:50).

5. Compare Jorge's argument with critical approaches arguing that computer-generated poetry aligns with postmodern aesthetic sensibilities in the context of which "meaning and significance ... are formed in the mind of the reader" and "are not completely dependent on the verbal material itself," which often "challenges intelligibility" (Funkhouser 2007:18). Scholars writing on interactive digital art (which requires the active participation of the user) have approached the question of its frequent lack of intelligible meaning from another angle. For example, while Simanowski acknowledges that "the lack of narrativity and meaning ... is ... an element of digital literature and arts," he also argues that "interactive [digital] art immerses the audience into the work and thus allows focusing on action and play rather than on interpretation" (Simanowski 2010:25; see also Hayles and Montfort 2012 for "computer interactive fiction," and Koenitz, Ferri, Haahr, Sezen, and Sezen 2015 for "interactive digital narrative"). In this framework, digital literature's frequent lack of intelligible meaning ceases to be a problem once it is determined that such a meaning has never been its goal to begin with.

6. Of course, inasmuch as the artist has to work with specific hardware and software, he or she too is always already limited by some sort of material infrastructure.

7. Gregory's comparison fundamentally problematizes the following description of programming: "[Anne] makes a simple working program and shapes it gradually by successive modifications. She starts with a single black bird. She makes it fly. She gives it color. Each step is a small modification to a working program that she has in hand. If a change does not work, she undoes it with another small change. She 'sculpts'" (Turkle and Papert 1990:139). Sculpting is precisely the opposite of the kind of programming described in this vignette.

8. The word "llama" appears a number of times in *Leaves of Grass* to denote not the animal, but rather a master or guru.

9. Recent attempts to design plot generators are not markedly different in essence from Calvino's formal exercise in creating an algorithm-based narrative that focuses first and foremost on logical coherence. Thus the architecture of MEXICA, a computer-based plot generator, integrates "preconditions and postconditions" that have to do with "logical" commonsense knowledge and "social" commonsense knowledge. An example of a precondition that focuses on logical common sense is "the action character A heals character B requires that character B is ill or injured in order for it to be performed ... Otherwise, it does not make sense to cure character B. This precondition does not depend on the social context" (Pérez y Pérez 2019:263). The postcondition of this action is the following: "The consequence of the action character A heals character B is that character B is healthy again (the tension health at risk is deactivated). It does not depend on cultural traditions" (264). Leaving aside the many context-specific and culturally specific reasons the implications of these conditions might not follow naturally or logically from

them, what this and similar exercises demonstrate is the extreme difficulty of formalizing common sense or situated action of any kind (Metz 2016; Suchman 2007).

10. Note that "the most distilled example of code poetry . . . [is] a form of homophonous poetry, that is, a translation from one language into another based on the sounds of the words, not their meaning or spelling" (Tomasula 2012:488). As I discussed in chapter 6, the practice of "homophonic translation"—of "translating for sound rather than sense"—has a "venerable history" among members of the Oulipo (Duncan 2019:22). For the prevalence of the haiku in the early days of digital poetry in general, see Funkhouser (2007:34–35, 56).

11. Of course, these two orientations can coexist even in the same individual as part of his or her "cultural repertoire" or "tool kit" (Swidler 1986).

12. To be sure, even computer-generated poetry's resulting textures and raised questions can be interpreted by the reader as making a rhetorical point (see, e.g., Hayles 2009:44–45).

13. There was nothing original about my intention or the computational means by which I tried to realize it. A somewhat similar intention motivated Christopher Strachey to write his groundbreaking 1952 computer-generated poem *Love Letters*, by means of which he tried to denaturalize "normative expressions of desire" and "words most socially marked as sincere in mainstream English society" (Wardrip-Fruin 2011:315).

Conclusion

1. One reason for the dystopian nature of many anthropologists' accounts of roboprocesses may be contemporary cultural anthropology's strong impulse to be critical of capitalism and the state and their presumed totalizing power, and to speak on behalf of their presumed suffering subjects (Robbins 2013) who are often conceptualized as powerless (e.g., Besteman 2019:171–174) or, at best, as being in possession of ad hoc resistance strategies in the form of "weapons of the weak" (Gusterson 2019:23). Although these accounts can, at times, be on the mark, they often tend to miss the affordances provided (rather than the vulnerabilities produced) by capitalism and the state, which individuals willingly heed in order to enrich their lives.

References

Abrams, M. H. 1971. *The Mirror and the Lamp: Romantic Theory and the Critical Tradition*. London: Oxford University Press.
Ake, David. 2002. "Jazz 'Training': John Coltrane and the Conservatory." In *Jazz Cultures*, pp. 112–145. Berkeley and Los Angeles: University of California Press.
Arnaud, Noel. 1998. "Foreword." In *Oulipo: A Primer of Potential Literature*, ed. Warren F. Motte, pp. xi–xv. Normal, IL: Dalkey Archive Press.
Bailey, Richard W. 1973. "Preface." In *Computer Poems*, ed. Richard W. Bailey. Drummond Island, MI: Potagannissing.
Bakhtin, Mikhail. 1982. "Discourse in the Novel." In *The Dialogic Imagination: Four Essays*, ed. Michael Holquist, trans. Caryl Emerson and Michael Holquist, pp. 259–422. Austin: University of Texas Press.
Balpe, Jean-Pierre. 2003. "E-Poetry: Time and Language Changes." Paper presented at "E-Poetry 2003: An International Digital Poetry Festival," West Virginia University, Morgantown, April 26.
Barthes, Roland. 1978. "The Death of the Author." In *Image, Music, Text*, trans. Stephen Heath, pp. 142–148. New York: Hill & Wang.
Barthes, Roland. 1982. *Camera Lucida: Reflections on Photography*. New York: Farrar, Straus, & Giroux.
Bauman, Richard. 1984. *Verbal Art as Performance*. Long Grove, IL: Waveland Press.
Bauman, Richard, and Charles L. Briggs. 2003. *Voices of Modernity: Language Ideologies and the Politics of Inequality*. Cambridge and New York: Cambridge University Press.
Becker, Gaylene. 1997. *Disrupted Lives: How People Create Meaning in a Chaotic World*. Berkeley and Los Angeles: University of California Press.
Becker, Howard S. 1982. *Art Worlds*. Berkeley and Los Angeles: University of California Press.
Bénabou, Marcel. 1998. "Rule and Constraint." In *Oulipo: A Primer of Potential Literature*, ed. Warren F. Motte, pp. 40–47. Normal, IL: Dalkey Archive Press.
Benjamin, Walter. 1969. "The Work of Art in the Age of Mechanical Reproduction." In *Illuminations*, ed. Hannah Arendt, pp. 217–252. New York: Schocken Books.
Bens, Jacques. 1998. "Queneau Oulipian." In *Oulipo: A Primer of Potential Literature*, ed. Warren F. Motte, pp. 65–73. Normal, IL: Dalkey Archive Press.

Berliner, Paul. 1994. *Thinking in Jazz: The Infinite Art of Improvisation.* Chicago: University of Chicago Press.

Berryman, Sylvia. 2007. "The Imitation of Life in Ancient Greek Philosophy." In *Genesis Redux: Essays in the History and Philosophy of Artificial Life,* ed. Jessica Riskin, pp. 35–45. Chicago: University of Chicago Press.

Besteman, Catherine, and Hugh Gusterson, eds. 2019. *Life by Algorithms: How Roboprocesses Are Remaking Our World.* Chicago: University of Chicago Press.

Besteman, Catherine. 2019. "Remaking the World." In *Life by Algorithms: How Roboprocesses Are Remaking Our World,* ed. Catherine Besteman and Hugh Gusterson, pp. 165–180. Chicago: University of Chicago Press, 2019.

Blanchette, Alex. 2019. "Infinite Proliferation, or The Making of the Modern Runt." In *Life by Algorithms: How Roboprocesses Are Remaking Our World,* ed. Catherine Besteman and Hugh Gusterson, pp. 91–106. Chicago: University of Chicago Press.

Block, Friedrich W., Christiane Heibach, and Karin Wenz, eds. 2004. *Poesis: The Aesthetics of Digital Poetry.* Ostfilden: Hatje Cantz Verlag.

Boden, Margaret. A. 2003. *The Creative Mind: Myths and Mechanisms.* London: Routledge.

Boellstorff, Tom. 2008. *Coming of Age in Second Life.* Princeton, NJ: Princeton University Press.

Bourdieu, Pierre. 1977. *Outline of a Theory of Practice.* Cambridge and New York: Cambridge University Press.

Bourdieu, Pierre. 1980. "The Aristocracy of Culture." *Media, Culture and Society* 2:225–254.

Bown, Oliver. 2014. "Empirically Grounding the Evaluation of Creative Systems: Incorporating Interaction Design." In *Proceedings of the Fifth International Conference on Computational Creativity,* ed. Simon Colton, Dan Ventura, Nada Lavrač, and Michael Cook, pp. 112–119. http://kt.ijs.si/publ/iccc_2014_proceedings.pdf, accessed November 28, 2022.

Boyer, Dominic. 2007. *Understanding Media: A Popular Philosophy.* Chicago: Prickly Paradigm.

Briggs, Charles L. and Richard Bauman. 1992. "Genre, Intertextuality, and Social Power." *Journal of Linguistic Anthropology* 2(2):131–172.

Brooks, Gwendolyn. 1969. *Riot.* Detroit: Broadside Press.

Callon, Michel, and John Law. 2005. "On Calculation, Agency, and Otherness." *Environment and Planning D: Society and Space* 23:717–733.

Calvino, Italo. 1998. "Prose and Anticombinatorics." In *Oulipo: A Primer of Potential Literature,* ed. Warren F. Motte, pp. 143–152. Normal, IL: Dalkey Archive Press.

Campbell, Colin. 1989. *The Romantic Ethic and the Spirit of Modern Consumerism.* Oxford: Blackwell.

Castle, Terry. 1995. *The Female Thermometer: Eighteenth-Century Culture and the Invention of the Uncanny.* Oxford and New York: Oxford University Press.

Chevigny, Paul. 2005 [1991]. *Gigs: Jazz and the Cabaret Laws in New York City.* New York: Routledge.

Cohen, Harold, Fride Nake, David C. Brown, Paul Brown, Philip Galanter, Jon McCormack, and Mark d'Inverno. 2012. "Evaluation of Creative Aesthetics." In *Computers and Creativity,* ed. Jon McCormack and Mark d'Inverno, pp. 95–114. Berlin: Springer.

Coleman, Gabriella E. 2010. "Ethnographic Approaches to Digital Media." *Annual Review of Anthropology* 39:487–505.

Coleman, Gabriella E. 2012. *Coding Freedom: The Ethics and Aesthetics of Hacking.* Princeton, NJ: Princeton University Press.

REFERENCES

Coleman, Gabriella E., and Alex Golub. 2008. "Hacker Practice." *Anthropological Theory* 8(3):255–277.
Collins, Nick, Alex McLean, Julin Rohrhuber, and Adrian Ward. 2003. "Live Coding in Laptop Performance." *Organized Sound* 8(3):321–329.
Cope, David. 2005. *Computer Models of Musical Creativity*. Cambridge, MA: MIT Press.
Csordas, Thomas J. 1993. "Somatic Modes of Attention." *Cultural Anthropology* 8 (2): 135–156.
Cukor, George, dir. 1964. *My Fair Lady*. 170 min. Warner Bros.
Derrida, Jacques. 1977. *Of Grammatology*, trans. Gayatri C. Spivak. Baltimore, MD: Johns Hopkins University Press.
Douglas, Mary. 1994. *Risk and Blame: Essays in Cultural Theory*. London: Routledge.
Downey, Gary L. 1998. *The Machine in Me: An Anthropologist Sits among Computer Engineers*. London: Routledge.
Du Bois, John W. 1993. "Meaning without Intention: Lessons from Divination." In *Responsibility and Evidence in Oral Discourse*, ed. Jane H. Hill and Judith T. Irvine, pp. 48–71. Cambridge and New York: Cambridge University Press.
du Sautoy, Marcus. 2019. *The Creativity Code: Art and Innovation in the Age of AI*. Cambridge, MA: Harvard University Press.
Duncan, Dennis. 2019. *The Oulipo and Modern Thought*. Oxford and New York: Oxford University Press.
Eastwood, Clint, dir. 1986. *'Round Midnight*. 133 min. Little Bear Productions.
Eckert, Penelope, and John R. Rickford, eds. 2002. *Style and Sociolinguistic Variation*. Cambridge and New York: Cambridge University Press.
Eisenlohr, Patrick. 2009. "Technologies of the Spirit: Devotional Islam, Sound Reproduction and the Dialectics of Mediation and Immediacy in Mauritius." *Anthropological Theory* 9(3):273–296.
Eisenlohr, Patrick. 2010. "Materialities of Entextualization: The Domestication of Sound Reproduction in Mauritian Muslim Devotional Practices." *Journal of Linguistic Anthropology* 20(2): 314–33.
Eubanks, Virginia. 2018. *Automating Inequality: How High-Tech Tools Profile, Police, and Punish the Poor*. New York: St. Martin's Press.
Evans-Pritchard, Edward E. 1991 [1937]. *Witchcraft, Oracles, and Magic Among the Azande*. London: Clarendon Press.
Faulkner, Robert R., and Howard S. Becker. 2009. *"Do You Know . . . ?" The Jazz Repertoire in Action*. Chicago: University of Chicago Press.
Fink, Robert. 2005. *Repeating Ourselves: American Minimal Music as Cultural Practice*. Berkeley and Los Angeles: University of California Press.
Finnegan, Ruth. 1988. *Literacy and Orality*. Oxford: Basil Blackwell.
Freud, Sigmund. 1961 [1928]. "Dostoevsky and Parricide." In *The Standard Edition of the Complete Psychological Works of Sigmund Freud*, vol. 21, *1927–1931*, ed. James Strachey, pp. 177–194. London: Hogarth.
Freud, Sigmund. 1964 [1919]. "The Uncanny." In *The Standard Edition of the Complete Psychological Works of Sigmund Freud*, vol. 22, *1917–1919*, ed. James Strachey, pp. 217–256. London: Hogarth.
Fried, Michael. 2005. "Barthes's Punctum." *Critical Inquiry* 31(3):539–574.
Funkhouser, Chris T. 2007. *Prehistoric Digital Poetry: An Archaeology of Forms, 1959–1995*. Tuscaloosa: University of Alabama Press.

Funkhouser, Christopher T. 2009. "Kissing the Steak: The Poetry of Text Generators." In *Literary Art in Digital Performance: Case Studies in New Media Art and Criticism*, ed. Francisco J. Ricardo, pp. 69–83. New York: Continuum.

Gal, Susan, and Judith T. Irvine. 2019. *Signs of Difference: Language and Ideology in Social Life*. Cambridge and New York: Cambridge University Press.

Garfinkel, Harold. 1967. *Studies in Ethnomethodology*. Cambridge: Polity.

Garfinkel, Harold. 2008. *Toward a Sociological Theory of Information*, ed. Anne W. Rawls. Boulder, CO: Paradigm Press.

Gates, Henry Louis, Jr. 1988. *The Signifying Monkey: A Theory of African-American Literary Criticism*. New York: Oxford University Press.

Gatti, Lorenzo, Gözde Özbal, Marco Guerini, Oliviero Stock, and Carlo Strapparava. 2019. "Computer-Supported Human Creativity and Human-Supported Computer Creativity in Language." In *Computational Creativity: The Philosophy and Engineering of Autonomously Creative Systems*, ed. Tony Veale and F. Amilcar Cardoso, pp. 237–254. Cham: Springer.

Gell, Alfred. 1994. "The Technology of Enchantment and the Enchantment of Technology." In *Anthropology, Art, and Aesthetics*, ed. Jeremy Coote and Anthony Shelton, pp. 40–63. Oxford: Clarendon Press.

Gendolla, Peter. 2009. "Artificial Poetry: On Aesthetic Perception in Computer-Aided Literature." In *Literary Art in Digital Performance: Case Studies in New Media Art and Criticism*, ed. Francisco J. Ricardo, pp. 167–177. New York: Continuum.

Geoghegan, Bernard D. 2011. "From Information Theory to French Theory: Jakobson, Lévi-Strauss, and the Cybernetic Apparatus." *Critical Inquiry* 38(1):96–126.

Gershon, Ilana. 2010. "Media Ideologies: An Introduction." *Journal of Linguistic Anthropology* 20(2):283–293.

Gershon, Ilana. 2011. "Neoliberal Agency." *Current Anthropology* 52(4):537–555.

Gershon, Ilana. 2017. "Language and the Newness of Media." *Annual Review of Anthropology* 46(15–31).

Gervas, Pablo. 2009. "Computational Approaches to Storytelling and Creativity." *AI Magazine* 30(3):49–62.

Gervas, Pablo. 2019. "Exploring Quantitative Evaluations of the Creativity of Automatic Poets." In *Computational Creativity: The Philosophy and Engineering of Autonomously Creative Systems*, ed. Tony Veale and F. Amilcar Cardoso, pp. 275–304. Cham: Springer.

Gioia, Ted. 1988. *The Imperfect Art*. Oxford and New York: Oxford University Press.

Givan, Benjamin. 2022. "How Democratic Is Jazz?" In *Finding Democracy in Music*, ed. Robert Adlington and Esteban Buch, pp. 58-79. London: Routledge.

Goffman, Erving. 1981. "Footing." In *Forms of Talk*, pp. 124–59. Philadelphia: University of Pennsylvania Press.

Grimstad, Paul. 2019. "The Absolute Originality of Georges Perec." *The New Yorker*, July 16, https://www.newyorker.com/books/page-turner/the-absolute-originality-of-georges-perec, accessed October 6, 2020.

Guilbaut, Serge. 1983. *How New York Stole the Idea of Modern Art: Abstract Expressionism, Freedom, and the Cold War*. Chicago: University of Chicago Press.

Gumperz, John J. 1992. "Contextualization and Understanding." In *Rethinking Context: Language as an Interactive Phenomenon*, ed. Alessandro Duranti and Charles Goodwin, pp. 229–252. Cambridge: Cambridge University Press.

REFERENCES

Gusterson, Hugh. 2019. "Robohumans." In *Life by Algorithms: How Roboprocesses Are Remaking Our World*, ed. Catherine Besteman and Hugh Gusterson, pp. 1–27. Chicago: University of Chicago Press.

Hastings, Adi, and Paul Manning. 2004. "Introduction: Acts of Alterity." *Language and Communication* 24:291–311.

Hayles, Katherine N. 1999. *How We Became Posthuman: Virtual Bodies in Cybernetics, Literature, and Informatics*. Chicago: University of Chicago Press.

Hayles, Katherine N. 2006a. "The Time of Digital Poetry: From Objet to Event." In *New Media Poetics: Contexts, Technotexts, and Theories*, ed. Adalaide Morris and Thomas Swiss, pp. 181–209. Cambridge, MA: MIT Press.

Hayles, Katherine N. 2006b. "Traumas of Code." *Critical Inquiry* 33(1):136–157.

Hayles, Katherine N. 2009. "Strickland and Lawson Jaramillo's *slippingglimpse*: Distributed Cognition at/in Work." In *Literary Art in Digital Performance: Case Studies in New Media Art and Criticism*, ed. Francisco J. Ricardo, pp. 39–51. New York: Continuum.

Hayles, Katherine N., and Nick Montfort. 2012. "Interactive Fiction." In *The Routledge Companion to Experimental Literature*, ed. Joe Bray, Alison Gibbons, and Brian McHale, pp. 452–466. London: Routledge.

Helmreich, Stefan. 1998a. "Recombination, Rationality, Reductionism and Romantic Reactions: Culture, Computers, and the Genetic Algorithm." *Social Studies of Science* 28(1):39–71.

Helmreich, Stefan. 1998b. *Silicon Second Nature: Culturing Artificial Life in a Digital World*. Berkeley and Los Angeles: University of California Press.

Helmreich, Stefan. 2003. "Trees and Seas of Information: Alien Kinship and the Biopolitics of Gene Transfer in Marine Biology and Biotechnology." *American Ethnologist* 30(3):340–358.

Hill, Jane H., and Judith T. Irvine, eds. 1993. *Responsibility and Evidence in Oral Discourse*. Cambridge and New York: Cambridge University Press.

Holland, John. 1992 [1975]. *Adaptation in Natural and Artificial Systems*. Cambridge, MA: MIT Press.

Hope, Cat, and John Ryan. 2014. *Digital Arts: An Introduction to New Media*. New York: Bloomsbury.

Ingold, Tim, and Elizabeth Hallam. 2007. "Creativity and Cultural Improvisation: An Introduction." In *Creativity and Cultural Improvisation*, ed. Elizabeth Hallam and Tim Ingold, pp. 1–24. London: Berg.

Irani, Lilly. 2015. "The Cultural Work of Microwork." *New Media and Society* 17(5):720–739.

Iversen, Margaret. 2010. "Introduction: The Aesthetics of Chance." In *Chance*, ed. Margaret Iversen, pp. 12-27. Cambridge, MA: MIT Press.

Jackson, Michael. 1989. *Paths Toward a Clearing: Radical Empiricism and Ethnographic Inquiry*. Bloomington: Indiana University Press.

Jakobson, Roman. 1960. "Concluding Statement: Linguistics and Poetics." In *Style in Language*, ed. Thomas A. Sebeok, pp. 350–377. Cambridge, MA: MIT Press.

Jentsch, Ernst. 1997 [1906]. "On the Psychology of the Uncanny." *Angelaki: Journal of the Theoretical Humanities* 2(1):7–16.

Jordanous, Anna. 2010. "A Fitness Function for Creativity in Jazz Improvisation and Beyond." In *Proceedings of the International Conference on Computational Creativity*, ed. Dan Ventura, Alison Pease, Rafael Pérez y Pérez, Graeme Ritchie, and Tony Veale, pp. 223–227. https://eden.dei.uc.pt/~amilcar/ftp/e-Proceedings_ICCC-X.pdf, accessed November 28, 2022.

Jordanous, Anna. 2019. "Evaluating Evaluation: Assessing Progress and Practices in Computational Creativity Research." In *Computational Creativity: The Philosophy and Engineering of Autonomously Creative Systems*, ed. Tony Veale and F. Amilcar Cardoso, pp. 211–236. Cham: Springer.

Kantosalo, Anna, Jukka M. Toivanen, and Hannu Toivonen. 2015. "Interaction Evaluation for Human-Computer Co-creativity: A Case Study." In *Proceedings of the Sixth International Conference on Computational Creativity*, ed. Hannu Toivonen, Simon Colton, Michael Cook, and Dan Ventura, pp. 276–283. https://computationalcreativity.net/iccc2015/proceedings/ICCC2015_proceedings.pdf, accessed November 28, 2022.

Katsuno, Hirofumi. 2011. "The Robot's Heart: Tinkering with Humanity and Intimacy in Robot-Building." *Japanese Studies* 31(1):93–109.

Keane, Webb. 1997. *Signs of Recognition: Powers and Hazards of Representation in an Indonesian Society*. Berkeley and Los Angeles: University of California Press.

Keane, Webb. 2007. *Christian Moderns: Freedom and Fetish in the Mission Encounter*. Berkeley and Los Angeles: University of California Press.

Kelty, Christopher. 2008. *Two Bits: The Cultural Significance of Free Software*. Durham, NC: Duke University Press.

Kiesling, Scott. 2009. "Style as Stance: Stance as the Explanation for Patterns of Sociolinguistic Variation." In *Stance: Sociolinguistic Perspectives*, ed. A. Jaffe, pp. 171–194. Oxford and New York: Oxford University Press.

Kockelman, Paul. 2016. *The Art of Interpretation in the Age of Computation*. Oxford and New York: Oxford University Press.

Koenitz, Hartmut, Gabriele Ferri, Mads Haahr, Diğdem Sezen, and Tognuç Ibrahim Sezen, eds. 2015. *Interactive Digital Narrative: History, Theory and Practice*. London: Routledge.

Koskimaa, Raine. 2010. "Approaches to Digital Literature: Temporal Dynamics and Cyborg Authors." In *Reading Moving Letters: Digital Literature in Research and Teaching*, ed. Roberto Simanowski, Joörgen Schäfer, and Peter Gendolla, pp. 129–143. Bielefeld: Transcript-Verlag.

Kunreuther, Laura. 2010. "Transparent Media: Radio, Voice, and Ideologies of Directedness in Postdemocratic Nepal." *Journal of Linguistic Anthropology* 20(2): 334–351.

Kurzweil, Ray. 2000. *The Age of Spiritual Machines: When Computers Exceed Human Intelligence*. London: Penguin.

Lamaison, Pierre, and Pierre Bourdieu. 1986. "From Rules to Strategies: An Interview with Pierre Bourdieu." *Cultural Anthropology* 1(1):110–20.

Landes, Joan B. 2007. "The Anatomy of Artificial Life: An Eighteenth-Century Perspective." In *Genesis Redux: Essays in the History and Philosophy of Artificial Life*, ed. Jessica Riskin, pp. 96–116. Chicago: University of Chicago Press.

Larkin, Brian. 2008. *Signal and Noise: Media, Infrastructure, and Urban Culture in Nigeria*. Durham, NC: Duke University Press.

Latour, Bruno. 2005. *Reassembling the Social*. Oxford and New York: Oxford University Press.

Lemon, Alaina. 2018. *Technologies for Intuition: Cold War Circles and Telepathic Rays*. Berkeley and Los Angeles: University of California Press.

Lescure, Jean. 1998. "Brief History of the Oulipo." In *Oulipo: A Primer of Potential Literature*, ed. Warren F. Motte, pp. 32–39. Normal, IL: Dalkey Archive Press.

Lewis, George E. 2000. "Too Many Notes: Computers, Complexity and Culture in Voyager." *Leonardo Music Journal* 10(1):33–39.

REFERENCES

Li, Tania M. 2007. *The Will to Improve: Governmentality, Development, and the Practice of Politics.* Durham, NC: Duke University Press.
Lord, Albert. 1960. *The Singer of Tales.* Cambridge, MA: Harvard University Press.
Lutz Fernandez, Anne, and Catherine Lutz. 2019. "Roboeducation." In *Life by Algorithms: How Roboprocesses Are Remaking Our World*, ed. Catherine Besteman and Hugh Gusterson, pp. 44–58. Chicago: University of Chicago Press.
MacIntyre, Alasdair C. 2007. *After Virtue: A Study in Moral Theory.* Notre Dame, IN: University of Notre Dame Press.
Mackenzie, Adrian. 2003. "Bringing Sequences to Life: How Bioinformatics Corporealizes Sequence Data." *New Genetics and Society* 22(3):315–332.
Madianou, Mirca, and Daniel Miller. 2013. "Polymedia: Towards a New Theory of Digital Media in Interpersonal Communication." *International Journal of Cultural Studies* 16(2):169–187.
Malaby, Thomas M. 2002. "Odds and Ends: Risk, Mortality, and the Politics of Contingency." *Culture, Medicine and Psychiatry* 26(3):283–312.
Malaby, Thomas M. 2007. "Beyond Play: A New Approach to Games." *Games and Culture* 2(2):95–113.
Manning, Paul. 2010. "Can the Avatars Speak?" *Journal of Linguistic Anthropology* 19(2):310–325.
Manning, Paul. 2018a. "Animating Virtual Worlds: Emergence and Ecological Animation of Ryzom's Living World of Atys." *First Monday* 23(6). https://firstmonday.org/ojs/index.php/fm/article/view/8127/7414, accessed November 28, 2022.
Manning, Paul. 2018b. "Spiritualist Signal and Theosophical Noise." *Journal of Linguistic Anthropology* 28(1):67–92.
Manovich, Lev. 2001. *The Language of New Media.* Cambridge, MA: MIT Press.
Manovich, Lev. 2003. "New Media from Borges to HTML." In *The New Media Reader*, ed. Noah Wardrip-Fruin and Nick Montfort, pp. 13–25. Cambridge, MA: MIT Press.
Masco, Joseph. 2019. "Ubiquitous Surveillance." In *Life by Algorithms: How Roboprocesses Are Remaking Our World*, ed. Catherine Besteman and Hugh Gusterson, pp. 125–144. Chicago: University of Chicago Press.
Mathews, Harry, and Alastair Brotchie, eds. 1998. *Oulipo Compendium.* London: Atlas.
Mayor, Adrienne. 2018. *Gods and Robots: Myths, Machines, and Ancient Dreams of Technology.* Princeton, NJ: Princeton University Press.
McCormack, Jon. 2012. "Creative Ecosystems." In *Computers and Creativity*, ed. Jon McCormack and Mark d'Inverno, pp. 39–60. Berlin: Springer.
McCormack, Jon. 2019. "Creative Systems: A Biological Perspective." In *Computational Creativity: The Philosophy and Engineering of Autonomously Creative Systems*, ed. Tony Veale and F. Amilcar Cardoso, pp. 327–352. Cham: Springer.
McCormack, Jon, and Mark d'Inverno, eds. 2012. *Computers and Creativity.* Berlin: Springer.
McGurl, Mark. 2009. *The Program Era: Postwar Fiction and the Rise of Creative Writing.* Cambridge, MA: Harvard University Press.
McLean, Alex, and Geraint Wiggins. 2012. "Computer Programming in the Creative Arts." In *Computers and Creativity*, ed. Jon McCormack and Mark d'Inverno, pp. 235–254. Berlin: Springer.
Merleau-Ponty, Maurice. 2002. *Phenomenology of Perception.* London: Routledge.
Metz, Cade. 2016. "One Genius' Lonely Crusade to Teach a Computer Common Sense." *Wired.* https://www.wired.com/2016/03/doug-lenat-artificial-intelligence-common-sense-engine/, accessed July 21, 2022.

Meyer, Birgit. 2006. "Impossible Representations: Pentecostalism, Vision, and Video Technology in Ghana." In *Religion, Media, and the Public Sphere*, ed. Birgit Meyer and Annelies Moors, pp. 290–312. Bloomington: Indiana University Press.

Miller, Brian A. 2020. "'All of the Rules of Jazz': Stylistic Models and Algorithmic Creativity in Human-Computer Improvisation." *MTO: Journal of the Society for Music Theory* 26(3).

Monson, Ingrid. 1996. *Saying Something: Jazz Improvisation and Interaction*. Chicago: University of Chicago Press.

Monson, Ingrid. 1998. "Oh Freedom: George Russell, John Coltrane, and Modal Jazz." In *In the Course of Performance: Studies in the World of Musical Improvisation*, ed. Bruno Nettl and Melinda Russell, pp. 149–168. Chicago: University of Chicago Press.

Monson, Ingrid. 2007. *Freedom Sounds: Civil Rights Call Out to Jazz and Africa*. Oxford and New York: Oxford University Press.

Montfort, Nick. 2007. "Narrative and Digital Media." In *The Cambridge Companion to Narrative*, ed. David Herman, pp. 172–186. Cambridge and New York: Cambridge University Press.

Montfort, Nick. 2015. "Foreword." In *Interactive Digital Narrative: History, Theory and Practice*, ed. Hartmut Koenitz, Gabriele Ferri, Mads Haahr, Diğdem Sezen, and Tognuç Ibrahim Sezen, pp. ix–xiii. London: Routledge.

Moore, Omar Khayyam. 1957. "Divination—A New Perspective." *American Anthropologist* 59(1):69–74.

Mori, Mashahiro. 2012. "The Uncanny Valley." *IEEE Robotics and Automation Magazine* 19(2): 98–100.

Morozov, Evgeny. 2013. *To Save Everything, Click Here: The Folly of Technological Solutionism*. New York: PublicAffairs.

Motte, Warren F. 1986. "Clinamen Redux." *Comparative Literature Studies* 23(4):263–281.

Motte, Warren F. 1998. "Introduction." In *Oulipo: A Primer of Potential Literature*, ed. Warren F. Motte, pp. 1–22. Normal, IL: Dalkey Archive Press.

Murray, Penelope. 1989. "Poetic Genius and Its Classical Origins." In *Genius: The History of an Idea*, ed. Penelope Murray, pp. 1–13. Oxford: Blackwell.

Nicholson, Stuart. 2005. "Teachers Teaching Teachers: Jazz Education." In *Is Jazz Dead? (Or Has It Moved to a New Address)*, pp. 99–127. London: Routledge.

Nierhaus, Gerhard. 2010. *Algorithmic Music Composition: Paradigms of Automated Music Generation*. New York: Springer.

Noble, Safiya U. 2018. *Algorithms of Oppression: How Search Engines Reinforce Racism*. New York: NYU Press.

Novak, David. 2013. *Japanoise: Music at the Edge of Circulation*. Durham, NC: Duke University Press.

Nozawa, Shunsuke. 2016. "Ensoulment and Effacement in Japanese Voice Acting." In *Media Convergence in Japan*, ed. Patrick W. Galbraith and Jason G. Karlin, pp. 169–199. N.p.: Kinema Club.

O'Neil, Cathy. 2016. *Weapons of Math Destruction: How Big Data Increases Inequality and Threatens Democracy*. New York: Broadway Books.

Olding, Rachel. 2019. "Why Is Jazz Unpopular? The Musicians 'Suck,' Says Branford Marsalis." *Sunday Morning Herald*, April 19. https://www.smh.com.au/entertainment/music/why-is-jazz-unpopular-the-musicians-suck-says-branford-marsalis-20190312-p513h2.html, accessed October 5, 2020.

Orr, Julian. 1996. *Talking about Machines: An Ethnography of a Modern Job*. Ithaca, NY: ILR.

Owens, Thomas. 2002. "Analyzing Jazz." In *The Cambridge Companion to Jazz*, ed. Marvyn Cooke and David Horn, pp. 286–297. Cambridge and New York: Cambridge University Press.
Pachet, François. 2002. "Playing with Virtual Musicians: The Continuator in Practice." *IEEE Multimedia* 9(3):77–82.
Pachet, François. 2003. "The Continuator: Musical Interaction with Style." *Journal of New Music Research* 32(3):333–341.
Pachet, François. 2012. "Musical Virtuosity and Creativity." In *Computers and Creativity*, ed. Jon McCormack and Mark d'Inverno, pp. 115–146. Berlin: Springer.
Pariser, Eli. 2011. *The Filter Bubble: What the Internet Is Hiding from You*. London: Viking.
Park, George K. 1963. "Divination and Its Social Contexts." *Journal of the Royal Anthropological Institute of Great Britain and Ireland* 93(2):195–209.
Peirce, Charles S. 1992. "On the Algebra of Logic: A Contribution to the Philosophy of Notation." In *The Essential Peirce*, vol. 1, ed. Nathan Houser and Christian Kloesel, pp. 225–228. Bloomington: Indiana University Press.
Perec, Georges. 1969. *La disparition*. Paris: Gallimard.
Pérez y Pérez, Rafael. 2019. "Representing Social Common-Sense Knowledge in MEXICA." In *Computational Creativity: The Philosophy and Engineering of Autonomously Creative Systems*, ed. Tony Veale and F. Amilcar Cardoso, pp. 255–274. Cham: Springer.
Perloff, Marjorie. 1991. *Radical Artifice: Writing Poetry in the Age of Media*. Chicago: University of Chicago Press.
Perloff, Marjorie. 2012. *Unoriginal Genius: Poetry by Other Means in the New Century*. Chicago: University of Chicago Press.
Petrusich, Amanda. 2017. "A Quest to Rename the Williamsburg Bridge for Sonny Rollins." *The New Yorker*, April 5. https://www.newyorker.com/culture/cultural-comment/a-quest-to-rename-the-williamsburg-bridge-for-sonny-rollins, accessed October 5, 2020.
Porter, Eric. 2002. *What Is This Thing Called Jazz: African American Musicians as Artists, Critics, and Activists*. Berkeley and Los Angeles: University of California Press.
Queneau, Raymond. 1998. "Potential Literature." In *Oulipo: A Primer of Potential Literature*, ed. Warren F. Motte, pp. 51–64. Normal, IL: Dalkey Archive Press.
Rawls, Anne W. 2008. "Introduction." In Harold Garfinkel, *Toward a Sociological Theory of Information*, ed. Anne W. Rawls, pp. 1–100. Boulder, CO: Paradigm Press.
Retallack, Joan, ed. 1996. *MUSICAGE: Cage Muses on Words, Art, Music*. Hanover, NH: Wesleyan University Press.
Richardson, Kathleen. 2016. "Technological Animism: The Uncanny Personhood of Humanoid Machines." *Social Analysis* 60(1):110–28.
Ritchie, Graeme. 2019. "The Evaluation of Creative Systems." In *Computational Creativity: The Philosophy and Engineering of Autonomously Creative Systems*, ed. Tony Veale and F. Amilcar Cardoso, pp. 159–194. Cham: Springer.
Robbins, Joel. 2013. "Beyond the Suffering Subject: Toward an Anthropology of the Good." *Journal of the Royal Anthropological Institute*, n.s., 19(3):447–462.
Robertson, Jennifer. 2018. *Robo Sapiens Japanicus: Robots, Gender, Family, and the Japanese Nation*. Berkeley and Los Angeles: University of California Press.
Rosaldo, Renato. 1986. "Ilongot Hunting as Story and Experience." In *The Anthropology of Experience*, ed. Victor W. Turner and Edward M. Bruner, pp. 97–138. Urbana: University of Illinois Press.

Rosenthal, David H. 1992. *Hard Bop: Jazz and Black Music, 1955–1965*. Oxford and New York: Oxford University Press.
Roubaud, Jacques. 1998. "Mathematics in the Method of Raymond Queneau." In *Oulipo: A Primer of Potential Literature*, ed. Warren F. Motte, pp. 79–96. Normal, IL: Dalkey Archive Press.
Royle, Nicholas. 2003. *The Uncanny*. Manchester: Manchester University Press.
Rubin, Rick. 2021. *Brian Eno: The Innovator*. Broken Record Podcast. https://www.youtube.com/watch?v=BOtrCYyf4cg, accessed July 19, 2022.
Saemmer, Alexandra. 2010. "Digital Literature—A Question of Style: Traditions and Approaches of Digital Literature in France: A Short Overview." In *Reading Moving Letters: Digital Literature in Research and Teaching*, ed. Roberto Simanowski, Jörgen Schäfer, and Peter Gendolla, pp. 163–182. Bielefeld: Transcript-Verlag.
Sahlins, Marshall D. 1985. *Islands of History*. Chicago: University of Chicago Press.
Sawyer, Keith. 1996. "The Semiotics of Improvisation: The Pragmatics of Musical and Verbal Performance." *Semiotica* 108: 269–306.
Sawyer, Keith. 2003. *Group Creativity: Music, Theater, Collaboration*. New York: Psychology Press.
Schegloff, Emanuel A., Gail Jefferson, and Harvey Sacks. 1977. "The Preference for Self-Correction in the Organization of Repair in Conversation." *Language* 53(2):361–382.
Schieffelin, Edward L. 1985. "Performance and the Cultural Construction of Reality." *American Ethnologist* 12(4):707–724.
Schüll, Natasha D. 2012. *Addiction by Design: Machine Gambling in Las Vegas*. Princeton, NJ: Princeton University Press.
Schüll, Natasha D. 2016. "Data for Life: Wearable Technology and the Design of Self-Care." *BioSocieties* 11(3):317–333.
Schuller, Gunther. 1999. "Sonny Rollins and the Challenge of Thematic Improvisation." In *Musings: The Worlds of Gunther Schuller*, pp. 86–97. New York: Da Capo Press.
Seaver, Nick. 2012. "Algorithmic Recommendations and Synaptic Functions." *limn* 2. http://limn.it/algorithmic-recommendations-and-synaptic-functions/, accessed October 23, 2022.
Shim, Eunmi. 2007. *Lennie Tristano: His Life in Music*. Ann Arbor: University of Michigan Press.
Silverstein, Michael. 2005. "Axes of Evals: Token versus Type Interdiscursivity." *Journal of Linguistic Anthropology* 15(1):6–22.
Silvio, Teri. 2010. "Animation: The New Performance?" *Journal of Linguistic Anthropology* 20(2):422–438.
Simanowski, Roberto. 2010. "Reading Digital Literature: A Subject Between Media and Methods." In *Reading Moving Letters: Digital Literature in Research and Teaching*, ed. Roberto Simanowski, Jörgen Schäfer, and Peter Gendolla, pp. 15–28. Bielefeld: Transcript-Verlag.
Simon, Herbert. 2019. *The Sciences of the Artificial*. Cambridge, MA: MIT Press.
Simonite, Tom. 2019. "A Health Care Algorithm Offered Less Care to Black Patients." *Wired*, October 24. https://www.wired.com/story/how-algorithm-favored-whites-over-blacks-health-care/, accessed October 7, 2020.
Spitulnik Vidali, Debra. 2010. "Millennial Encounters with Mainstream Television News: Excess, Void, and Points of Engagements." *Journal of Linguistic Anthropology* 20(2):372–88.
Stacey, Jackie, and Lucy Suchman. 2012. "Animation and Automation: The Liveliness and Labours of Bodies and Machines." *Body and Society* 18(1):1–46.

Sterne, Jonathan. 2002. *The Audible Past: Cultural Origins of Sound Reproduction*. Durham, NC: Duke University Press.

Stout, Noelle. 2019. "Automated Expulsion in the U.S. Foreclosure Epidemic." In *Life by Algorithms: How Roboprocesses Are Remaking Our World*, ed. Catherine Besteman and Hugh Gusterson, pp. 31–43. Chicago: University of Chicago Press.

Strickland, Stephanie. 2006. "Writing the Virtual: Eleven Dimensions of E-Poetry." *Leonardo Electronic Almanac* 14(5). https://www.leoalmanac.org/wp-content/uploads/2012/09/06Writing-the-Virtual-Eleven-Dimensions-of-E-Poetry-by-Stephanie-Strickland-Vol-14-No-5-6-September-2006-Leonardo-Electronic-Almanac.pdf, accessed October 6, 2020.

Swidler, Ann. 1986. "Culture in Action: Symbols and Strategies." *Annual Review of Sociology* 51(2):273–286.

Suchman, Lucy. 2007. *Human-Machine Reconfigurations: Plans and Situated Actions* [2nd ed.]. Cambridge and New York: Cambridge University Press.

Suchman, Lucy. 2012. "Don't Kick the Dog." In *Robot Futures: Lucy Suchman's Reflections on Technocultures of Humanlike Machines*. https://robotfutures.wordpress.com/2012/09/12/dont-kick-the-dog/, accessed October 5, 2020.

Taussig, Karen-Sue, Klaus Hoeyer, and Stefan Helmreich. 2013. "The Anthropology of Potentiality in Biomedicine." *Current Anthropology* 54(7):S3–S14.

Taussig, Michael. 1992. *Mimesis and Alterity*. London: Routledge.

Taylor, Charles. 1989. *Sources of the Self: The Making of the Modern Identity*. Cambridge, MA: Harvard University Press.

Taylor, Grant D. 2014. *When the Machine Made Art: The Troubled History of Computer Art*. New York: Bloomsbury.

Terrio, Susan J. 2019. "Detention and Deportation of Minors in U.S. Immigration Custody." In *Life by Algorithms: How Roboprocesses Are Remaking Our World*, ed. Catherine Besteman and Hugh Gusterson, pp. 59–76. Chicago: University of Chicago Press.

Thurman, Neil, Seth C. Lewis, and Jessica Kunert, eds. 2019. "Algorithms, Automation, and News," special issue, *Digital Journalism* 7(8).

Tomasula, Steve. 2012. "Code Poetry and New-Media Literature." In *The Routledge Companion to Experimental Literature*, ed. Joe Bray, Alison Gibbons, and Brian McHale, pp. 483–496. London: Routledge.

Turkle, Sherry, 1991. "Romantic Reactions: Paradoxical Responses to the Computer Presence." In *Boundaries of Humanity: Humans, Animals, Machines*, ed. James J. Sheehan and Morton Sosna, pp. 224–252. Berkeley and Los Angeles: University of California Press.

Turkle, Sherry. 2011. *Alone Together: Why We Expect More from Technology and Less from Each Other*. New York: Basic Books.

Turkle, Sherry, and Seymour Papert. 1990. "Epistemological Pluralism: Styles and Voices within the Computer Culture." *Signs* 16(1):128–157.

Turner, Fred. 2006. *From Counterculture to Cyberculture: Stewart Brand, the Whole Earth Network, and the Rise of Digital Utopianism*. Chicago: University of Chicago Press.

Veale, Tony, and F. Amilcar Cardoso, eds. 2019. *Computational Creativity: The Philosophy and Engineering of Autonomously Creative Systems*. Cham: Springer.

Ventura, Dan. 2019. "Autonomous Intentionality in Computationally Creative Systems." In *Computational Creativity: The Philosophy and Engineering of Autonomously Creative Systems*, ed. Tony Veale and F. Amilcar Cardoso, pp. 49–70. Cham: Springer.

Vidler, Anthony. 1996. *The Architectural Uncanny: Essays in the Modern Unhomely.* Cambridge, MA: MIT Press.

Walser, Robert. 1997. " 'Out of Notes': Signification, Interpretation, and the Problem of Miles Davis." In *Keeping Score: Music, Disciplinarity, Culture,* ed. David Schwartz, Anahid Kassabian, and Lawrence Siegel, pp. 147–168. Charlottesville: University Press of Virginia.

Wardrip-Fruin, Noah. 2010. "Five Elements of Digital Literature." In *Reading Moving Letters: Digital Literature in Research and Teaching,* ed. Roberto Simanowski, Jörgen Schäfer, and Peter Gendolla, pp. 29–57. Bielefeld: Transcript-Verlag.

Wardrip-Fruin, Noah. 2011. "Digital Media Archaeology: Interpreting Computational Processes." In *Media Archaeology: Approaches, Applications, and Implications,* ed. Erkki Huhtamo and Jussi Parikka, pp. 302–322. Berkeley and Los Angeles: University of California Press.

Webster, Anthony K. 2015. "Cultural Poetics (Ethnopoetics)." In *Oxford Handbooks Online.* https://www.oxfordhandbooks.com/view/10.1093/oxfordhb/9780199935345.001.0001/oxfordhb-9780199935345-e-34.

Webster, Anthony K. 2020. "Learning to be Satisfied: Navajo Poetics, a Chattering Chipmunk, and Ethnopoetics." *Oral Tradition* 34:73–104.

Weidman, Amanda J. 2003. "Guru and Gramophone: Fantasies of Fidelity and Modern Technologies of the Real." *Public Culture* 15(3):453–476.

Weidman, Amanda J. 2010. "Sound and the City: Mimicry and Media in South India." *Journal of Linguistic Anthropology* 20(2):294–313.

Weizenbaum, Joseph. 1966. "ELIZA—A Computer Program For the Study of Natural Language Communication between Man and Machine." *Computational Linguistics* 9(1):36–45.

Wiener, Norbert. 1954 [1950]. *The Human Use of Human Beings: Cybernetics and Society.* Garden City, NY: Doubleday.

Wiener, Norbert. 1982 [1948]. *Cybernetics: Or Control and Communication in the Animal and the Machine.* Cambridge, MA: MIT Press.

Wiggins, Geraint A. 2019. "A Framework for Description, Analysis and Comparison of Creative Systems." In *Computational Creativity: The Philosophy and Engineering of Autonomously Creative Systems,* ed. Tony Veale and F. Amilcar Cardoso, pp. 21–47. Cham: Springer.

Wilce, James M. 2001. "Divining *Troubles,* or Divining Troubles?: Emergent and Conflictual Dimensions of Bangladeshi Divination." *Anthropological Quarterly* 74(4):190–200.

Wilf, Eitan. 2010. "Swinging within the Iron Cage: Modernity, Creativity, and Embodied Practice in American Postsecondary Jazz Education." *American Ethnologist* 37(3):563–582.

Wilf, Eitan. 2011. "Sincerity versus Self-Expression: Modern Creative Agency and the Materiality of Semiotic Forms." *Cultural Anthropology* 26(3):462–484.

Wilf, Eitan. 2012. "Rituals of Creativity: Tradition, Modernity, and the 'Acoustic Unconscious' in a U. S. Collegiate Jazz Music Program." *American Anthropologist* 114(1):32–44.

Wilf, Eitan. 2013a. "From Media Technologies that Reproduce Seconds to Media Technologies that Reproduce Thirds: A Peircean Perspective on Stylistic Fidelity and Style-Reproducing Computerized Algorithms." *Signs and Society* 2(1):185–211.

Wilf, Eitan. 2013b. "Sociable Robots, Jazz Music, and Divination: Contingency as a Cultural Resource for Negotiating Problems of Intentionality." *American Ethnologist* 40(4):605–618.

Wilf, Eitan. 2013c. "Streamlining the Muse: Creative Agency and the Reconfiguration of Charismatic Education as Professional Training in Israeli Poetry-Writing Workshops." *Ethos* 41(2):127–149.

Wilf, Eitan. 2013d. "Toward an Anthropology of Computer-Mediated, Algorithmic Forms of Sociality." *Current Anthropology* 54(6):716–739.
Wilf, Eitan. 2014a. *School for Cool: The Academic Jazz Program and the Paradox of Institutionalized Creativity.* Chicago: University of Chicago Press.
Wilf, Eitan. 2014b. "Semiotic Dimensions of Creativity." *Annual Review of Anthropology* 43:397–412.
Wilf, Eitan. 2019. *Creativity on Demand: The Dilemmas of Innovation in an Accelerated Age.* Chicago: University of Chicago Press.
Wilf, Eitan. 2020. "'The Closed World Principle': Corporations and the Metaculture of Newness via Oldness." *Journal of Business Anthropology* 9(1):1–18.
Wilson, Olly. 1999. "The Heterogeneous Sound Ideal in African-American Music." In *Signifyin(g), Sanctifyin(g), & Slam Dunking: A Reader in African American Expressive Culture*, ed. Gena Caponi, pp. 157–171. Amherst: University of Massachusetts Press.
Wilson, Robert N. 1986. *Experiencing Creativity: On the Social Psychology of Art.* New Brunswick, NJ: Transaction Books.
Wise, Norton M. 2007. "The Gender of Automata in Victorian Britain." In *Genesis Redux: Essays in the History and Philosophy of Artificial Life*, ed. Jessica Riskin, pp. 163–194. Chicago: University of Chicago Press.
Woodmansee, Martha. 1996. "Genius and the Copyright." In *The Author, Art, and the Market: Rereading the History of Aesthetics*, pp. 35–56. New York: Columbia University Press.
Young, Bob. 2006. "Branford Marsalis Sounds Off on All That Jazz." *Boston Herald*, September 11.
Young, Michael, and Tim Blackwell. 2016. "Live Algorithms for Music: Can Computers be Improvisers?" in *The Oxford Handbook of Critical Improvisation Studies*, vol. 2, ed. Benjamin Piekut and George E. Lewis, pp. 507–528. Oxford and New York: Oxford University Press.
Zeitlyn, David. 2012. "Divinatory Logics: Diagnoses and Predictions Mediating Outcomes." *Current Anthropology* 53(5):525–546.

Index

abstract expressionism, 19, 178
afk (away from keyboard), as fundamental to cybersociality, 127–28
Africa, 17
agency, 9, 93, 192, 200
"aha effects," 97–98
Alamo (group), 236n2
algorithmic self, 223
Algorithms of Oppression (Noble), 168
All the Things You Are (tune), 72–73, 76
alterity, 65, 125, 130
Amazon, 221
ambient music, 151–52
American liberal subjectivity, 178–80; and anthropology, 235n2; computational indeterminacy and conflicting cultural currents that inform, 161–62
animation, 115, 127, 141, 165; theory of, 125
anthropology, 125, 129, 162, 228n17, 233n3, 239n1; American liberal subjectivity, as informed by, 235n2
Apple Macintosh PowerBook 180, 57
artificial intelligence, 2, 36–37, 130–31, 198
artificial life, 42; "postcreation events" in, 43–44
Association for the Advancement of Creative Musicians (AACM), 232n16
authorial agency, and social justice, 171
authorial intentionality, 161
automatic writing, 192–93
Automating Inequality (Eubanks), 168
automation, 5, 7, 220–21
automatons, 119, 125; and watchmaking technology, 6–7
avant-garde, 123, 161, 178–80
Azande, 17–18

Bacon, Francis, 121
Barthes, Roland, 200–201, 238n4; punctum, 211; studium, 211–12
Bateson, Gregory, 116, 129, 228n17
Beckett, Samuel, 145
Beethoven, Ludwig van, 91
Bell, Alexander Graham, 95
Benedict, Ruth, 228n17
Benjamin, Walter, 145–46, 203–5
Bens, Jacques, 191
Berge, Claude, 191
Berklee College of Music, 9–10, 12, 88
Billie's Bounce (tune), 10, 227n11
bioinformatics, 232n5
"Black Perl" (code poem), 215
Blake, William, 173
blues, 10, 12, 72, 82
Boas, Franz, 228n17, 228n21
Boden, Margaret, 26, 89, 228n23
Booker's Waltz (tune), 61
Boston Dynamics, BigDog robot, 230n3
"bothunting," 234n10
Boulez, Pierre, 232n3
Bourdieu, Pierre, 228n17; on distance from necessity, 179; habitus, notion of, 11–12, 228n16
Bown, Oliver, 228n24
Braffort, Paul, 236n2
Brooks, Gwendolyn, 213
Brown, Clifford, 35, 42, 229n1
"Burning of the Abominable House, The" (Calvino), 214

Cage, John, 19–20, 236n8
Calliope (electronic poet), 201
Calvino, Italo, 191, 213–14, 236n4, 238n9
Carpenter, Jim, 154–55

Castle, The (Kafka), 147–48
chance, 16–17, 192–93, 211, 237n9; aesthetics of, 19–20
Chaplin, Charlie, 232n1
Charade (tune), 61
Chicago (Illinois), 213
Chopin, Frédéric, 92, 127
clinamen, 189–90, 237n6
coding, 5, 28–29, 35, 139; creative expression, as informed by Romantic tropes of, 165–66; indeterminacy of, 166; live, 165
Cold War, 178–79
Coltrane, John, 1, 75, 94–95, 98–100, 227n4, 231nn12–13
common sense, 71, 220–21, 238n9
communism, 178
computational creativity, 24, 26, 81, 228nn23–24; problems of evaluation in, 42–43, 45, 48–51, 132, 229nn6–7, 230n9, 230n11, 233nn8–9
computational indeterminacy, 161; and creative intentionality, 180; and potentiality, 29, 135
computer-generated jazz, 30, 31
computer-generated poetry, 8, 139, 200, 224, 234n3, 235n10, 238n5, 239n12; and American liberal subjectivity, 162, 178, 180; and "antmakers," 195; as art form, 147; code for, writing of, 134–35, 142–55, 157, 162–65, 168–69, 172, 180, 183–86, 194–95, 199, 205–6, 209, 211, 216; collaborative nature of, 185; as combinatoric, 142; computational indeterminacy in, and authorial intentionality, 161; computational indeterminacy of outcomes in, 147, 169–71; computational indeterminacy of process in, 147, 161–69, 173; and "computer agency," reification of, 173–75; constraints in, invention of, 142–45, 183, 203–7; contingency in, 30; and creative intentionality, 169, 180; as creative practice, 158, 171, 180; and creativity as spontaneous force, 190–91; as critical practice, 213; as digital literature, 141; as digital poetry, subgenre of, 141–42; exclusion in, 179, 236n11; and Google Translate, 140–41, 234n2; ideas for, 142–44; as inclusive kind of literary practice, 185; and intentionality, 162–69, 213–14; linguistic textures of, 152–53, 216–17; meaning in, 177–78; modification and improvement of, 145–46, 148; narrative development in, absence of, 151; and nondigital poetry, 29, 155–56, 164, 199, 207–8; novelty of, 177; Oulipo's influence on, 183–84, 236n11; output texts, 140–42, 146–51, 153, 155–56, 159–60, 164, 169, 175, 177, 179–80, 184, 194–95, 207–12, 216, 218, 235n6; "perfect poems," 209; permutational procedures of, 142; poetic function of, 176; as precise form of art, 199; and proximal multiplicity, 145–46; public reading of, 157; punctum of, 211–12; pure contingency in, 178; in Python 3 programming language, 140–41, 143–44, 147–48, 155, 175, 184; *qua* texture, 156; randomization in, 4, 147, 150, 152, 160–61, 169, 171–72, 176–77, 179, 182–83, 190–91, 211; referential function of, 175–76, 180; repetition in, 153–54, 176–77, 199, 212, 235n10, 235n13; sampling in, 150–51, 157, 207; as self-reflexive practice, 170–71; and self-tracking technologies, 170–71; semiotic contingency in, 178; and social inequality, 213; socialization into, 146–154; and spontaneity, 163, 166, 168, 175; style in, 144–45; substance in, lack of, 177; as syntactic templates, 142; text-distributional aesthetics of, 151–52, 154–56, 172, 177–78, 186, 208, 216–17, 234n8; use of hardware in, 143, 198; "verbage" of, 156
Computer History Museum, 57–58, 230n12
computer poems, and chance, 237n9
computer programming, 144, 153–54
context insensitivity, 232n15
contextualization cues, 73
contingency, 16–17, 30–31, 64; in art production, 19–20; of digital computation, 20; and intentionality, 18–19
Continuator, The, 86, 90, 97–98
conversation analysis: other-correction in, 115; "repairs" in, 115–16
copyright, 14
Corea, Chick, 86, 90–91, 99, 232n2
Costa Rica, 47
creative agency, 4, 8, 12, 15, 23, 93, 133, 143, 173; of computers, 174, 194; and digital computation, 16; and feeling, notion of, 13; and mimetic capacity, 7; and modernity, 5–6; Romantic form of, 180
creative computing, 167
creative contexts, 24
creative intentionality, 16, 29, 135, 161; and computational indeterminacy, 180; and digital computation, 20; as self-determination, 180
creative practice, computation-based, 165, 175
creativity, 6–7, 12, 15, 19, 21, 26, 39–40, 44, 49–50, 83, 85, 131–33, 167; aided by computer, 191; as algorithmic calculation, 101–5; in computational systems, 229n7; computerized systems, simulated by, 134; and digital computation, 24; ideological status of, 5; kinds of, 89; Romantic notions of, 23, 146, 161, 165–66, 168, 173–75, 235n4, 235n7; as spontaneous force, 190; as "thing," 134
criminal justice, 4–5
cybernetics, 118, 232n1, 233n3; and homeostasis, 114, 116, 129; and negative feedback mechanisms, 117
cybersociality, afk (away from keyboard) as fundamental to, 127–28

INDEX

Dadaism, 19, 183, 236n1; randomness in, 176
David's system, 29, 83, 85, 87, 103–5, 108, 126, 134, 224, 230n8, 230nn11–12, 230n7, 232n16; and academic jazz education, negative consequences of, 79–80; and "aha effects," 97–98; and audience indifference, 109; and "cheating," notion of, 46; chord-scale theory, reliance on, 80; computational architecture, maintaining of, 88; and conceptual spaces, already existing, 89; context sensitivity, lack of, 75, 81; control of, 41, 45, 54, 62; embodied form, lack of, 109; execution of pieces by, 62–63; genetic algorithms in, 39–40, 46; harmonic knowledge of, 63–64; improvisation in, 8–9, 27, 36, 38–39, 48, 51, 54, 67–68, 70–71, 73–76, 79–80, 86, 89, 133, 220, 230n9, 231nn13–14; and jam sessions, 35, 41–42, 44, 55, 62, 68–69; and licks, 44–45; limitations of, 70–78, 231n13; measure populations in, 43, 47–48; as microworld, 42–44, 70; as modest, 133; musical interaction with, 65–70; mutation operators in, 40–41, 46, 49–50, 52, 54, 57, 67–68, 79, 220; as narcissistic experience, 64–66, 133, 222; performance, control of, 88; performative contingency in, 64; phrase populations in, 40, 45–46, 48; pure contingency in, 64; repair work, inability to do, 74; repertoire of, 61–62; semiotic contingency in, 64; as sideman, 55–56; standardized quality of, 77; stylistic predictability of, 54–55, 60–61; synthesized sound of, 77; trading fours in, 65–70; uniform random generator, use of, 44. *See also* jazz-improvising digital systems
Davis, Miles, 1–3, 60, 74–75, 87, 90–91, 93–94, 98–99, 104, 227n4; reinventing himself, 88
Days of Wine and Roses (tune), 62
"Death of the Author, The" (Barthes), 200
defamiliarization, 123
Derrida, Jacques, on metaphysics of presence, 163
Descartes, René, 160, 174, 180
differentiations, 163
digital art: and narrative, 234n7; randomness in, 235n5; and sculpture, 206–7
digital computation, 7, 17, 83–84, 134, 141, 158, 184; Commodore 64, 197–98; contingency in, cultivation of, 20–22; and creative agency, 16; and creative intentionality, 20; in creative practice, 24–25, 171; for creative purposes, 21, 28–29; as creative tool, 23; and creativity, 3, 24; human creativity enriched by, 24; and intentionality, 16; in jazz and poetry, 31; as nonhuman agent, 4, 24; for poetic purposes, 183, 198, 223; transformation from enabling means to limiting end, 197–98; writing and editing code for, 29, 205
digital computational architecture, 85–86
digital gambling machines, 65; and random number generators (RNG), 172–73

digital media, 25
digital poetry, 25, 143, 148, 167, 184, 236n1; randomization in, 171, 176
digital reproduction, 205
digital technologies, 3–4, 6–8, 15–16, 19, 21–27, 31–32, 110, 166–67, 204; and customized content, 221–22; and filter bubble, 221–22
Disappearance, The (Perec), 188
divination, 18; mechanical, 17, 20
DOCTOR program, 116
Dorham, Kenny, 10, 14
Dostoevsky, Fyodor, 121
Duchamp, Marcel, 19, 183, 236n1
Ducrocq, Albert, 201

Edison, Thomas, 95–96
Eisenlohr, Patrick, 8
Electronic Text Composition, 154
ELIZA programs, 116
embodied forms, 28, 88, 105, 109–11, 126, 131, 234n8
Enlightenment, 5, 114, 121–22, 129, 233n4
Eno, Brian, 151–52, 158
entropy, 116
ethnopoetics, 29
ethnopsychology, 113–14
Europe, 6–7, 13
Evans-Pritchard, Edward E., and poison oracle, 17–18
evolution, 37, 47–48
evolutionary-computation framework, 56–57
exclusion, aesthetics of, 179, 236n11
expert knowledge, 21–23

Facebook, 221–22
fascism, 178
Fichte, Johann Gottlieb, 14
film studies, 125
fixed texts, 8–9, 15
Four (tune), 52, 71, 74
France, 56
free will, 17
Freud, Sigmund, 121, 123; on uncanny, 119–20, 122
Fried, Michael, 212
Funkhouser, Christopher, 25

gambling, 66; and repeated play, 65
Garfinkel, Harold, 123, 233nn1–3, 233n7; on continuity, disruption of, 119; on deprivation of secondary information, 128; ethnomethodological theory, 115–17; ethnomethodological theory, cybernetic roots of, 114, 116–17, 129; on incongruity, notion of, 117; on incongruity as noise, 117–18; on interactional homeostasis, 118; on noise as disruption, 119–21; orchestrated experiments of, 118; on social order, cybernetic

Garfinkel, Harold (*cont.*)
  roots of, 117–18; thermostat example, 117; on zero-order information, 119, 121, 128
Geertz, Clifford, 228n17
Gell, Alfred, 228n17
generation zero, 37
generative music systems, 31
genetic algorithms, 30, 44, 46–50, 101, 155, 220; computational framework, 27, 37, 39, 133, 229n5; and crossover points, 52–53; and fitness function, 41, 45, 52, 56, 72
Genetic and Evolutionary Computation Conference (GECCO), 56–57
Giant Steps (Coltrane), 78, 231nn12–13
Gillespie, Dizzy, 92
Goblin Market (Rossetti), 209
Goffman, Erving, 65
Google, 221–22
Google Translate, 140–41, 175, 234n11
Gordon, Dexter, 11
Gusterson, Hugh, 220

habitus, 11–12, 228n16, 228n18
Hayles, Katherine, 25, 230n10, 232n1; on metaphorical slippage, 229n4
Helmreich, Stefan, 42–43, 47, 188
Hoeyer, Klaus, 188
Hoffmann, E. T. A., 122
Holland, John, schema theorem, 229n5
homeostasis, 114; interactional, 117–18, 129–30
House of Dust, A (Knowles and Tenney), 167, 169; code of, 213
Hubbard, Freddie, 10, 227n13
human-machine interaction, 59, 70, 114–15, 118–19, 129–30; as corrective to disappointment with fellow humans, 64–65; studies of, emerging from cybernetics, 116

I Ching, 20
I Dream of Jeannie (tune), 61
imitation, 11
Impressions (tune), 75
improvisation, 220–21; bebop, 229n4, 231n13; in David's system, 8–9, 27, 36, 38–39, 48, 51, 54, 67–68, 70–71, 73–76, 79–80, 86, 133, 230n9, 231n13; in James's system, 86–87, 90, 97, 100, 133; in jazz, 3, 10–11, 16, 28, 31, 36, 38–40, 51, 54, 67–68, 70–71, 73–76, 79–80, 86–87, 90–91, 94, 97, 99–100, 104, 126, 228n15, 229n3, 230n9, 230n17, 231n13; joint, 36, 51, 74, 105–6, 126, 133; of licks, 48, 54; real time of, 70, 86; regulated, 228n16; slow, 231n13; of Syrus, 3, 97, 99–100, 104–7, 126, 133; as unique, 39–40
incongruity, as noise, 117–18
indeterminacy, 20, 161; of outcomes, 169, 174–75; of process, 147, 163, 166, 173–75

Industrial Revolution, 85, 232n1
information technologies, 129
information theory, 129
intentionality, 4, 15, 17, 21–22, 113, 161, 213–14, 223; and contingency, 18–19; and digital computation, 16
interactional noise, 114, 118, 127; aesthetic pleasure, as source of, 130
intertextuality, 8, 122, 213

Jakobson, Roman, 129; on poetic function, 176
James's system, 98–99, 103–5, 111, 115, 118, 124–25, 132–34, 224, 230n11; acoustic sound of, 88; and conceptual spaces, new, 89; generative algorithms, 89; and ideas in new styles, 89; improvisation in, new styles of, 90–91, 97, 100–101, 220–21; jazz musicians sounding alike as reason for developing, 84–87, 102; mirroring of musicians' styles, critique of, 87–88; mixing styles in, 92–94; noise, 121; perceptual algorithms in, 89. *See also* jazz-improvising digital systems; Syrus
jam sessions, 60–62, 64, 69, 73; trading fours during, 66–68
Japanese noise-music scene, and positive feedback loops, 233n3, 235n3
jazz, 12, 29, 101; in academic training, 9, 22, 27, 31, 36, 56, 58, 79–82, 84–85, 88, 102–3, 224, 232n16, 235n2; chord-scale theory, 38, 80; "coming out of the cold," 69; context sensitivity in, 75; contextualization cues in, 73; cultural order of, 35, 56, 69, 77, 83; emergent nature of, 73; flexibility in, 74–75, 78, 110, 130; as hierarchical, 56; improvisation, 37–38, 70, 75–76, 94, 228n15, 229n3, 230n7, 231n13; institutionalization of, 31; interactivity of, 70, 76, 81; jam sessions, 61–62; licks, 9, 36, 38, 42, 51–52, 54–55, 61; motivic development in, 70, 80–81; musical sociality in, 58; quoting in, 70–71; sounding the same, 84–85; style-mixing in, 92–93; swinging in, 76; as "think-tank music," 102; trading fours in, 67; as unique and identifiable, 76–77
jazz-improvising digital systems, 30, 36–38, 50, 59, 62; academic jazz education, reproducing ill effects of, 79–81; algorithmic architecture of, 67, 71; and artificial algorithms, 55; and audience indifference, risk of, 81–82, 109; and ballads, 78; chord-scale theory, reliance on, 80; as competent sideman, 55–56; context sensitivity, lack of, 75–76, 78, 81, 131; contingency in, 64; crossover points in, 52–53; fitness function in, 41, 45, 52, 56, 72; genetic algorithms in, 39, 56, 58; harmonic knowledge of, 63; and human-generated notes, 77; and human voice, 83; and human withdrawal, 81; and improvisations, 55–56, 60, 63, 67, 71–73, 76, 86; and interaction,

55–57; and jam sessions, 60–61, 64, 66–69, 79; licks in, 61, 72, 76; masking, 70; motivic development in, 71–72; phrase populations in, 51–52, 54; pitch-to-MIDI interface of, 51; pragmatic inflexibility in, 76; repair work in, 74; rhythm section of, 74–76, 78; trading fours in, 67–69, 73, 231n9; unique sound, lack of, 76–77. *See also* David's system; James's system
Jentsch, Ernst, 120; on epilepsy as form of noise, 121

Kafka, Franz, 147–48
Kant, Immanuel, 179
Keane, Webb, 18
Kismet robot, 113
Knowles, Alison, 152, 167
Kroeber, Alfred, 228n17
Kurzweil, Ray, 47

Lacan, Jacques, 176; on unconscious as language, 187–88
language ideology, 163
La Vie mode d'emploi (Perec), 237n6
Leaves of Grass (Whitman), 209; "llama" as word, in, 210, 238n8
Le Lionnais, François, 187
Lescure, Jean, 176
Lévi-Strauss, Claude, 129; on "science of the concrete," 176
Lewis, George, and design of Voyager, 232n16
Li, Tania, 22–23
liberal scholarship, 167–68
liberal subjectivity, 174
"Life Design," 228n22
Locke, John, 121–22
Louis XVI, 6
Love Letters (Strachey), 152, 234n9 (chap. 5), 239n13
Lucretius, notion of clinamen, 189
Lutz, Theo, 147–48, 150, 175, 179
Lynch, David, 123

machine learning, 4, 134, 174–75
MacIntyre, Alasdair, 18–19
Macy Conferences, 129
Maillardet's automaton, 6–7
Mallarmé, Stéphane, 19
Manovich, Lev, 25, 144, 229n3
Marie Antoinette, 6
Marsalis, Branford, 102–3
"Mary Had a Little Lamb" (song), 96
Massachusetts Institute of Technology (MIT), Kismet robot, 113
materiality, 113, 130, 133; of icons and indices, 233n4; of semiotic forms, 114, 121, 124; in virtual worlds, 128

Matisse, Henri, 92
McLuhan, Marshall, 123–24
Mead, Margaret, 129
media technologies, 9, 11
mediation, 11, 16, 28, 94, 125–26, 130, 221; of immediation, 230n12; of Syrus, 96–97; visual worlds, as basis of, 127–28
metaphorical slippage, 38, 48, 229n4
MEXICA, 238n9
Miller, G. A., 116
Milton (Blake), 173
mimicry, 29
Mingus, Charles, 71
Moberly, L. G., 123
Moby-Dick (Melville), 155
modernity, 31–32; and creativity agency, 5–6
Modern Times (film), 232n1
Monk, Thelonious, 94, 98–100
Montfort, Nick, 215–16
"Moonlight" Sonata (Beethoven), 91–92
Morozov, Evgeny, 22
Mozart, Wolfgang Amadeus, 92
Music of Changes (Cage), 20
My Fair Lady (film), 228

National Science Foundation, 84
natural selection, 38
neoliberal agency, 171
Neroli (Eno), 151
new media, 25–27
new-media poetry, 217
New York City, 2, 28, 139, 196, 236n11
Noble, Safiya, 168
nondigital creative-writing programs and workshops, 31, 167–68
nondigital poetry, 195, 199, 207–11
Novak, David, 233n3

One Hundred Thousand Billion poems (Queneau), 186–88, 194, 203
originality, 5–6, 13–15, 19, 29, 190
Oulipo, 30, 142–45, 180, 183, 195, 198, 213–14, 217, 234n3, 235n4, 236n8 (chap. 6), 236nn1–2 (chap. 7), 236n4 (chap. 7), 238n4; on aid of computer and creativity, 191; chance, aversion to, 192–93; and clinamen, notion of, 189–91, 237n6; constraints, use of, 204–5; and homophonic translation, 176, 239n10; isolated genius, aversion to, 185; mathematically informed kinds of constraints, use of, 184, 187, 190, 199–200, 203; philosophy of, 236n3; playfulness of, 215; on potentiality, 189, 193–94; potentiality, fetishization of, 186–87, 236n4; "potential literature" of, 188; Romantic notion of poetic inspiration, rejection of, 184–85, 190, 192; S + 7 writing method of, 192–93, 204; on self-fashioned constraints,

Oulipo (cont.)
185–87; Snowball method of, 204; structuralism and Lacanian psychoanalysis, appropriation of, 176; and structuralist zeitgeist, 200; and surrealism, 192, 200–201; "synthesis," principle of, 184

Pachet, François, 86; Flow Machine, 232n3; on slow improvisation, 231n13; Virtuoso jazz-improvising system, 229n4, 230n7, 231n11
Parker, Charlie, 1, 3, 227n2, 228n15, 231n8, 232n3
participation framework, 8, 15
Perec, Georges, 142–45, 188–89, 204, 236n1, 237n6
Perloff, Marjorie, 167
phenomenology, 18
Picasso, Pablo, 92
Pietism, 13
Poe, Edgar Allan, 122
poetry, 28–29, 31, 41, 208; computer, use of in, 191
poison oracle, 17–18
politics of mediation, 8
potentiality, 29, 124, 186, 188–89, 237n5; as double-edged sword, 194; poetic, 135; as uncertain, 191–93
pragmatism, 18
Princeton University, Organizational Behavior Project, 116
Project Gutenberg, 209
Protestant Pietism, 13
psychiatry, 116
public sphere, 221–22

Queneau, Raymond, 185–88, 192–95, 199, 203, 214–15

randomization, 16, 152, 169, 171, 173; in computer-generated poetry, 160–61, 191; in "computer thought," 161; and digital gambling machines, 172
rationalization, 5, 80–81, 84
Raven, The (Poe), 210
Ray, Tom, 48; and Tierra, 47
Riot (Brooks), 213
roboprocesses, 5, 221, 224; in banking system, 220; and deskilling, 223; dystopic assessments of, 222, 239n1; and interpersonal relationships, zombification of, 222; proprietary knowledge, based on, 219
robots, 59; alterity, as lacking, 65, 130; companionship, 65–66, 130, 133; improvisation of, 84; as sociable, 65–66, 70, 112–13, 130, 133
Roentgen, David, 6
Rollins, Sonny, 231n8
Romanticism, 5–6, 13, 16, 23, 114, 122; creative artist, notion of, 193; creativity, notions of, 235n7; literature, 119, 123; poetic inspiration, notion of, 184, 213; self, notion of, 223
Roubaud, Jacques, 189, 199, 236n2, 237n6

'Round Midnight (film), 11–12, 14
Ruesch, Jurgen, 116

Sahlins, Marshall, on double contingency, 18
sampling, 150–51, 157, 207
"Sandman, The" (Hoffmann), 123
Sapir, Edward, 228n17
Schelling, Friedrich, 122
Schüll, Natasha Dow, 65–66
science and technology studies, 125
Sciences of the Artificial, The (Simon), 227n5
Second Life, afk (away from keyboard) in, 127–28
self-determination, 179; and creative intentionality, 180
semiotic ideology, 233n4; materiality in, 121–24; and noise, 122–23, 129; and self, materiality of forms linked to, 121
Sentimentalism, 13
Shannon, Claude, 116, 196–98
Silicon Valley, 22
Simanowski, Roberto, 238n5
Simon, Herbert, 195, 202, 227n5
sociality, 25, 29, 58, 83, 128, 133, 223–24; alterity, 125; animation, 125
social justice, 162, 168; and authorial agency, 171
sociology, 233n3
solutionism, 22–23
sound fidelity, 9, 15, 93–94, 97
sound reproduction, 95; and "aha effects," 97–98; sound fidelity in, 93–94, 97; stylistic fidelity in, 93–94
So What (tune), 75
speech, 95
Stacey, Jackie, 125
Stanislavski, Konstantin, 233n7
Sterne, Jonathan, 94–95, 97
Stochastic Texts (Lutz), 147, 152, 155, 157, 175, 179
Strachey, Christopher, 152, 239n13
structuralism, 176
style, as ideology, 12–14
stylistic fidelity, 9, 15, 28, 93–94, 96
Suchman, Lucy, 112, 125, 230n3; Garfinkel's influence on, 114, 116; work of, as revolutionary, 116
surrealism, 19, 181, 183, 192–93, 195, 200–201
Syrus, 1–3, 8, 27–28, 84, 91, 93–94, 115, 220, 222, 232n8; as acoustic instrument-playing robotic musician, 105; aesthetic pleasure in form of, 111; and "aha effects," 97–98; algorithmic architecture of, 89–90, 96, 99–100; arms of, 105, 124–25, 127; audience, ability to engage, 108–9; classical style or jazz style, responses in, 92; digital computational architecture, based on, 85; as embodied form, 105, 107–8, 110–11, 126, 130–33, 234n8 (chap. 4); "far out" playing, 104; and fractal-based algorithms, 101; and generative algorithms, 89; head of, 108–9, 112, 126–27;

head of, bobbing of, 126; and higher-order sequences, 96; human interaction with, 131–33; improvisation of, 96, 105–7, 126; interactional inflexibility of, 127; interactional noise of, 127; jazz master's corpus as database for, 95–96; listeners, difficulty relating to, 101–4, 109, 111; and lower-order sequences, 96; Markov models for, 86, 89–90, 96, 98, 101; materiality of, 124–25, 130, 133; MIDI format of solos in database for, 90; mixing of styles, 97–100; and "Motor music!" exclamation, 221; and noise, 129; own style of playing, 110, 130; "Parkinson moment," 112–14, 118, 120, 124; and perceptual algorithms, 89; playing "in the style of," 98; priming of audience, 99–100; repair work, inability to perform, 127, 221; and self-expression, illusion of, 126–27; and "superstyle," attempt to create, 99–100; and "Syrus music," 110, 130; as technology of enchantment, 109; as technology of estrangement, 127; as uncanny, 124; as underwhelming for team members, 99; unique style of, 110, 130. *See also* James's system

Taussig, Karen-Sue, 188
Taylor, Grant D., 230n10
Tenney, James, 152, 167
Turing, Alan, 196–98, 234n10
Turkle, Sherry, 65–66, 113, 222–23; on "robotic moment," 133
Turner, Fred, 234n11
Turner, Victor, 228n21
Twin Peaks (television series), 123
type intertextuality, 8–9

uncanny, 28, 114–15, 119–21, 125, 128; as aesthetic pleasure, 123; anthropological definition of, 124, 233n5; haunted houses as, 122–23; and materiality of semiotic forms, awareness of, 122; and noise, 123, 129; as Romantic phenomenon, 122–23
United States, 1, 8, 27–28, 35, 84, 168, 234n11; academic jazz programs in, 9, 22, 36, 85, 232n16; jazz training in, 36
Unoriginal Genius (Perloff), 167
Urdu poetry, 8

Vian, Boris, 201
video games, and child psychology, 65
virtual worlds: aesthetic pleasure in, 128; avatars in, 128; and mediation, 127–28; sociality in, 128
visual arts, 28–29

Walter, William Grey, Elmer and Elsie robots ("Grey Walter's tortoises"), 59, 230n2
Weber, Max, 228n21
Weizenbaum, Joseph, 36–37; ELIZA programs, 112–13, 116
Whitman, Walt, 209–10
Wiener, Norbert, 116, 232n1; thermostat example, 117
"Work of Art in the Age of Mechanical Reproduction, The" (Benjamin), 203
World War II, 129, 178

Yardbird Suite (tune), 1, 227n2
YouTube, 221

Printed and bound by CPI Group (UK) Ltd, Croydon, CR0 4YY
09/06/2025

14685759-0002